Innovation in Risk Analysis

Series Editor

Jianping Li, School of Economics and Management, University of Chinese Academy of Science, Beijing, China

This book series focuses on discussing the new theories, new models, and new methods of risk analysis. Moreover, the better risk management practices in the areas of social, physical and health sciences, engineering, public policy and administration, and media and communication studies will priority attention.

The book series aims to publish the latest theoretical and empirical research on the communication, regulation, and management of risk. Research that you might want to contribute to the book series could explore:

- The Inter-relationships between risk, decision-making and society.
- How to promote better risk management practices.
- Contribute to the development of risk management methodologies in the different areas.

More information about this series at https://link.springer.com/bookseries/16914

Jianping Li · Lu Wei · Xiaoqian Zhu

Financial Statements-Based Bank Risk Aggregation

Springer

Jianping Li
School of Economics and Management
University of Chinese Academy of Sciences
Beijing, China

Xiaoqian Zhu
School of Economics and Management
University of Chinese Academy of Sciences
Beijing, China

Lu Wei
School of Management Science
and Engineering
Central University of Finance
and Economics
Beijing, China

This book is supported by grants from the National Natural Science Foundation of China (92046023, 71425002, 71971207, 72001223).

ISSN 2731-6254 ISSN 2731-6262 (electronic)
Innovation in Risk Analysis
ISBN 978-981-19-0407-3 ISBN 978-981-19-0408-0 (eBook)
https://doi.org/10.1007/978-981-19-0408-0

This Springer imprint is published by the registered company Springer Nature Singapore Pte Ltd.
The registered company address is: 152 Beach Road, #21-01/04 Gateway East, Singapore 189721, Singapore

Contents

Chapter 1
Basic Concepts of Bank Risk Aggregation

1.1 The Basel Accords and Bank Risks

The bank is one of the most important financial institutions in the economic system. The stability of the bank is crucial to the stability of the whole financial system and the smooth operation of the economy (Allen and Faff 2012). A strong and resilient banking system is the foundation for sustainable economic growth, as banks are at the centre of the credit intermediation process between savers and investors. Moreover, banks provide critical services to consumers, small and medium-sized enterprises, large corporate firms and governments who rely on them to conduct their daily business, both at a domestic and international level.

In the process of operation, banks will inevitably face various types of risks, such as credit risk, market risk and operational risk (Basel Committee on Banking Supervision, BCBS for short 2006). The core of ensuring the soundness of commercial banks lies in the ability to manage risks (Berger et al. 2016).

In the 1970s, financial markets became increasingly globalized, and international finance and speculative activities prevailed. In the process of continuous internationalization, international commercial banks have become more and more separated from domestic banking supervision, and the joint supervision of international banks is very weak, causing great loopholes in bank supervision. Since that time, there have been major financial accidents caused by various risks.

In 1986, the Minneapolis First System Bank of the United States caused substantial losses to the bank due to errors in the prediction of interest rate trends. In 1995, the Bank of Barings bought a large number of futures contracts due to negligence, which led to the bankruptcy of the leading British banking industry. Also in 1995, the operator of Japan's Daiwa Bank bought and sold US bonds off-book, and eventually suffered a large loss. In 2001, Enron, an energy company located in Houston, Texas, went bankrupt within a few weeks due to financial fraud. The terrorist attacks of September 11 in 2001 caused economic losses of 200 billion U.S. dollars to the United States. Similarly, there are countless international incidents with credit risk, operational risk, and market risk.

J. Li et al., *Financial Statements-Based Bank Risk Aggregation*, Innovation in Risk Analysis, https://doi.org/10.1007/978-981-19-0408-0_1

A decision that seems inadvertent may cause the bank to fail. The results of these risks will not only affect domestic development and cause hundreds of millions of losses, but also have a chain reaction in the international banking industry, resulting in immeasurable and serious consequences. In 1974, three very famous international commercial banks—German Herstatt Bank, New York Franklin National Bank, and London's British Israel Bank collapsed one after another, causing huge losses to customers in many countries.

Under such circumstances, in 1975, with the initiation and support of the Bank for International Settlements, the 'Group of Ten' and the governors of the central banks of Switzerland and Luxembourg held a meeting in Basel, Switzerland, to discuss the international supervision and management of multinational banks. The permanent establishment of the "Basel Committee on Banking Supervision" was established. It is composed of senior representatives of banking supervisory agencies and central banks of Belgium, Germany, Canada and other countries.

The purpose of the committee is to strengthen cooperation among member states in bank supervision. It does not have any regulatory power over the state, nor does it have legal effect on member states. The Basel Committee on Banking Supervision is to promote international exchanges and cooperation among supervisory authorities of various countries, establish minimum supervisory standards that can be recognized in various fields, exchange financial supervision information, establish minimum supervisory standards that can be recognized in various fields, and maintain the stable operation of the international banking system.

In the early 1980s, international risks related to the debt crisis continued to increase, and the capital adequacy ratio of international banks also showed a downward trend. In order to ensure fair competition among international commercial banks, form a consistent standard for measuring capital adequacy, and prevent the decline of capital adequacy ratio, the Basel Committee formulated Basel Accord I, *International Convergence of Capital Measurement and Capital Standards*. Basel I provides the basic norms for the operation of international banks and stipulates the calculation standards for risk weights. It considers the credit risk of commercial banks in depth, but ignores market risk, operational risk, etc. Therefore, independent credit risk makes it impossible to consider the risk aggregation formed by multiple risks.

After the 1990s, under the continuous promotion of financial innovation, with the rapid growth of financial derivatives and transactions, the banking industry gradually intervened in derivatives transactions. Therefore, the volatility of the financial market has an increasing impact on the banking industry. In 1997, the Southeast Asian financial crisis broke out. From the financial turmoil, it is discovered that the problems in the financial industry are not only single risks such as credit risk or market risk, but are formed by the combined effect of credit risk, market risk, and operational risk.

However, in the early stage, relevant scholars believed that these risks were not related to each other, so they were measured separately and managed separately, and the economic capital of various risk types was estimated separately (BCBS 2009). However, the various risks of banks are not independent of each other, and there is a complex correlation between them (Imbierowicz and Rauch 2014). Their role in fueling the flames may even trigger a financial crisis and seriously threaten the stability of the financial system. During the subprime mortgage crisis, many European and American banks went bankrupt. With the poor management of bank operators, the traditional risk management model has been unable to meet the needs of bank risk management under the new situation, focusing on single risk and ignoring the correlation of multiple risks. Bank risk aggregation has gradually become the focus of academic and industry.

In this context, the Basel Committee decided to increase the risk sensitivity of the rules of the accord, and finally implemented the new Basel Capital Accord in 2004. The new Basel accord not only regulates credit risk and market risk, but also covers operational risk. The new accord inherits the old accord with capital adequacy as the core idea. The minimum capital adequacy ratio should reach 8%, and the bank's core capital adequacy ratio should be 4%. This can enhance the bank's sensitivity to risks and make operations more effective.

It is noteworthy that Basel II has effectively promoted banks' integrated risk management capabilities, encouraged attention to the relationship between risks, and improved the level of risk assessment. In modern life, as the types of risks gradually increase, banks are exposed to more and more risks, and many risks will inevitably influence each other. To strengthen the ability to identify risks, the Basel Committee on Banking Supervision believes that capital regulations and modern risk management should be combined, and risks should be emphasized through the disclosure of related risks and capital information.

With the implementation of Basel II and the continuous emergence of financial innovation, the status and management methods of Basel II are also constantly being challenged. After the outbreak of the financial crisis in 2007, Basel II exposed some problems, such as the inability to correctly measure the magnitude and relevance of risks, which led to the deterioration of the ability to resist risks. These problems show that there are still problems with the Basel Committee's regulatory thinking, indicating that capital's coverage and risk constraints cannot be equivalently converted. Therefore, in response to the financial crisis, the Basel Committee has formulated a number of documents to update the banking supervision mechanism, and these documents have formed Basel III.

It is not so much a replacement; it is better to say that Basel III is a supplement to Basel II. Basel III improves the risk coverage of the capital framework and once again emphasizes the risk integration by considering complex relationship between multiple risks, and thereby improves the quality of risk management. Specifically, besides traditional credit, market and operational risk, Basel III puts forward the 17 principles of liquidity risk to emphasize the management of liquidity risk. This document, together with the document Basel III: International framework for liquidity risk measurement, standards and monitoring, presents the Basel Committee's reforms to strengthen global capital and liquidity rules with the goal of promoting a more resilient banking sector.

What's more, Basel III once again emphasizes the relationship between multiple risks. In addition, it has also increased the requirements for information disclosure, disclosing various types of information and risks. Given the scope and speed with which the recent and previous crises have been transmitted around the globe as well as the unpredictable nature of future crises, the Committee is introducing a number of fundamental reforms to the international regulatory framework. The reforms strengthen bank-level, or micro-prudential, regulation, which will help raise the resilience of individual banking institutions to periods of stress. The reforms also have a macroprudential focus, addressing system-wide risks that can build up across the banking sector.

1.2 Bank Risk Types

Banks' business activities will inevitably produce a variety of risks, such as credit, market and operational risks (BCBS 2006; Imbierowicz and Rauch 2014; Beltrame et al. 2018). Specifically, credit, market and operational risks have been covered by Basel I and Basel II. Recently, liquidity risk has drawn increasing attention in both industry and academia, and has been covered under Basel III (BCBS 2010). Thus, credit, market, liquidity and operational risks are main risks faced by banks.

Since the creation of the Basel Accord, a total of three accords have been issued. In 1988, the Basel Accord I—*International convergence of capital measurement and capital standards*, was released. The Basel II—*International convergence of capital measurement and capital standards: a revised framework* was released in 2004. Basel III was released in 2010—*Basel III: A global regulatory framework for more resilient banks and banking systems.*

As shown in Fig. 1.1, the figure introduces the process of the introduction of three types of risks: credit risk, market risk, and operational risk into the Basel Accord. The 1988 Basel Accord I introduced credit risk, and in 1992 the Basel Accord I issued in 1988 was implemented. In the following years, two types of risks have been introduced into the Basel Accord. The market risk was introduced in 1996. Later, it was discovered that operational risk is also one of the risks that cannot be ignored.

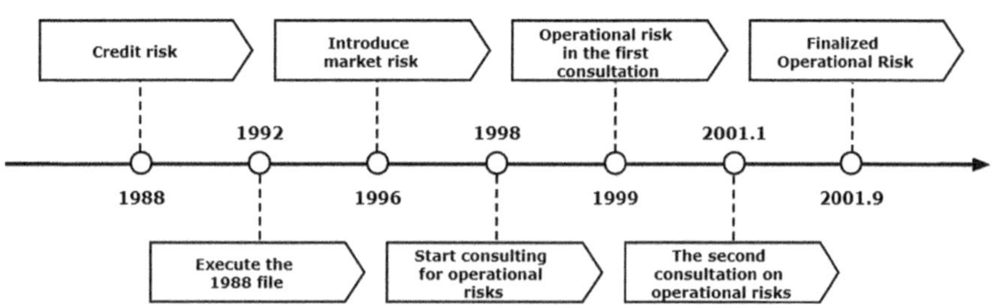

Fig. 1.1 The introduction process of the three main types of risks

The Basel Committee initiated the first and second consultations on operational risks in 1998, 1999, and 2001, and successfully finalized the operational risk in September 2001. Therefore, the three main types of risks are not introduced at the same time.

The first risk introduced in the Basel Capital Accord by BCBS is credit risk. It is widely known and considered to be one of the most important bank risks (Li et al. 2015). Basel norms define credit risk as the risk of counterparty failure, i.e. the risk of a loss caused by failure of the counterparty to fulfill its contractual obligations (BCBS 1988; Mustika et al. 2015). Credit risk not only refers to the potential loss from counterparty failure, but also includes loss caused by changes in the market value of the debt due to changes in the borrower's credit rating and performance capabilities. The Standardized Approach, the Foundation Internal Rating-based Approach and the Advanced Internal Rating-based Approach was proposed in Basel II for credit risk management (BCBS 2006).

The US subprime mortgage crisis is a typical example of credit risk, which has brought huge losses to banks and the entire financial system. This is a financial turmoil that occurred in the United States, due to the bankruptcy of a large number of subprime mortgage institutions, the forced closure of investment funds, and the violent volatility of the stock market. In 2006, the real estate market was sluggish, and the demand for housing gradually decreased. The owner of the property originally wanted to repay the loan by selling the house or mortgaged the house. However, the continued low temperature in the market has directly caused borrowers to be unable to repay loans from lending companies and banks, and these people are also burdened with large amounts of mortgage loans.

Eventually, it triggered a subprime mortgage crisis that brought serious disasters to the US financial market. According to the Federal Deposit Insurance Corporation, the subprime mortgage crisis resulted in 25 bank failures across the United States in 2008, 140 bank failures in 2009 and 157 bank failures in 2010. The core reason is that these people with very low credit ratings and no fixed income even bear huge debts, which brings huge credit risks that they may not be able to repay.

Although the Basel I formulated in 1988 does not cover market risk, people have been aware of market risk for a long time. Many banks mis-predicted interest rate and stock price fluctuations, causing them to face market risks and bring huge losses. The concept of market risk firstly appeared in the 1996 *Capital Accord Market Risk Supplementary Regulations*, which is a supplement of Basel I. Market risk is defined as the risk of losses in bank's on-balance sheet and off-balance sheet businesses due to adverse changes in market prices (interest rates, exchange rates, stock prices, and commodity prices). Market risk refers to the risk of losses on-and off-balance sheet positions, arising from movements in market prices, in Basel II Amendment (BCBS 2006).

In the mid-1980s, Minneapolis First System Bank predicted that interest rates would fall in the future, so it purchased a large number of government bonds. In 1986, interest rates fell as expected, which brought a lot of book gains. Unfortunately, in 1987 and 1988, interest rates continued to rise, and bond prices fell, causing the bank to lose up to 500 million U.S. dollars and eventually had to sell its headquarters building. Similarly, traders at Societe Generale in France caused huge losses due to illegal European stock index futures trading. Mis-predicting the trend of interest rates and violent stock price fluctuations cause banks to face market risks, which will bring huge losses.

In September 1998, the Basel Committee on Banking Supervision issued the *Operational Risk Management* for the first time, which included operational risk into the three major risks of the New Basel Accord. This is the operational risk, the risk of loss caused by imperfect or problematic internal procedures, personnel and systems, or external events. Operational risk is defined (under Basel Accord) as the risk of loss resulting from inadequate or failed internal processes, people and systems or from external events, which include legal risk but exclude strategic and reputational risk (BCBS 2006).

The bankruptcy of Barings Bank in 1995 embodied operational risks. Since the second half of 1994, Li Sen, the general manager and chief trader of Singapore Bahrain Futures Co., Ltd., has conducted a very complex, high-expectation, and extremely risky derivative financial commodity transaction in Tokyo Nikkei 225 Index Futures. Later, the Nikkei Index fell from January 1995, causing Li Sen's long position to suffer a heavy blow. To turn defeat and win, he bet on the rise of the Nikkei 225 index with a gambler mentality and continued to transfer huge amounts of money from London to buy a large number of futures contracts ultimately failed miserably.

On February 26, 1995, the Bahrain Bank, the leader of the British banking industry, with a glorious history of 233 years, was forced to declare bankruptcy due to a huge loss of 916 million pounds caused by Lisson, and ABN AMRO bank acquired it at a symbolic price of 1 pound. Internationally, there are also many major operational risks. For example, in 1995, Yamato Bank's management was chaotic, which caused a capital loss of 1.1 billion US dollars; in 2002, the National Australia Bank created false transactions to cover up losses caused by misjudgments in foreign exchange options transactions. A huge loss of 360 million Australian dollars. It shows that operational risks cannot be ignored.

Basel Committee on Banking Supervision issued *Principles for Sound Liquidity Risk Management and Supervision* in 2008, and defined liquidity risk as the ability of a bank to fund increases in assets and meet obligations as they become due, without incurring unacceptable losses (BCBS 2008). Basel III puts forward the principles of liquidity risk management in the second pillar, based on improving the risk coverage of the first pillar and confirming credit, market and operational risks. Although commercial banks have solvency, they cannot obtain sufficient funds in a timely manner, or cannot obtain sufficient funds promptly at a reasonable cost to cope with asset growth or pay due debts. The risks formed above are liquidity risks. The management of liquidity risk can not only appropriately control the stock to meet the daily operating needs of the bank, but also adjust the flow promptly to ensure the relative balance of the capital stock.

The North Rock Bank run in the United Kingdom is an example of liquidity risk. North Rock Bank is one of the most important housing mortgage banks in the UK. Its business model is to provide customers with a variety of loans. These loans can be secured or unsecured. At the same time, banks raise funds by attracting deposits, interbank lending, and mortgage asset securitization, and invest in bond markets outside the European market. US subordinated debt is also one of its important investment methods. However, after the subprime mortgage crisis in the United States, no bank was willing to provide funds to North Rock Bank. North Rock Bank has insufficient positions and can only seek help from the Central Bank (Bank of England). This caused the investors and depositors of Beiyan Bank to lose confidence. The stock price fell by nearly 80% in just a few trading days, and the UK's first run in 140 years occurred. Serious customer runs led to an outflow of more than 3 billion pounds, and the bank's total deposits were only 240 pounds. Therefore, this subprime mortgage crisis highlights the importance of liquidity risk management.

Nowadays, a total of eight bank risks are included in Basel III, namely credit risk, market risk, operational risk, liquidity risk, legal risk, country risk, reputation risk, and strategic risk (BCBS 2010). In daily business activities, if the bank violates relevant business standards and legal requirements, it will lead to the inability to perform the contract, litigation or other legal disputes. Eventually, this leads to economic losses of the bank, which is the legal risk. Inaccurate storage and handling of bank information lead to leakage of customer information, or banks privately inquiring about customer information, which will bring legal risks.

Country risk means that economic entities may suffer losses due to economic, political, and social changes in other countries, during international economic and financial transactions with non-residents of the country. In 1982, the governments of Mexico and Brazil announced the extension of debt repayment to Western creditor countries. Such default behaviors such as refusal to pay debts, delayed payment, and insolvency caused by the sovereignty of the country are related to national risk.

Reputational risk refers to the risk of a commercial bank's operation, management and external events, that will cause stakeholders to negatively evaluate the commercial bank. During 2008, the Bank of Halifax in Scotland encountered malicious rumors, which led to a sharp drop in the bank's stock price. In the Internet age, news spreads fast and spreads widely, and the reputation of commercial banks is facing challenges almost every day. Regarding strategic risks, in the process of systematic management of banks pursuing short-term business goals and long-term development goals, inappropriate future development plans and strategic decisions may threaten the bank's future development. Strategic risk is a factor that affects the development direction, corporate culture, information and survivability of the entire enterprise, or corporate benefits.

1.3 Bank Risk Aggregation

In 1995, a piece of news awakened the entire financial market. With a history of more than 230 years, the Bank of Barings of the United Kingdom, ranked 489th in terms of core capital among the world's 1000 largest banks, caused a huge loss of 916 million pounds due to speculative transactions in huge financial futures. After the National Central Bank, the Bank of England, failed to rescue over the previous weekend, it was forced to declare bankruptcy. In the above, we have mentioned the Bahrain Bank incident. This is a large-scale banking accident caused by multiple risks at the same time.

On the surface, Lisen's illegal operations and mis-judgment were the fuse of the entire incident. From an in-depth analysis, the high risk of financial derivatives is an important reason for this accident. From Lisson's personal mis-judgment to the bankruptcy of the entire Bank of Bahrain, the rate of return on investment has been multiplied because of financial derivatives, and the investment risk has also been multiplied. This is determined by the "leverage" characteristic of financial derivatives. The above is the result of the combined effect of operational risk and market risk. Therefore, in the actual environment, there are very few accidents with only one risk. Usually, risks are in the form of clusters.

The lack of a specialized risk management mechanism is an important reason why Lee Sen of the Bank of Bahrain can successfully engage in ultra vires transactions. Banks must establish a special risk management mechanism to deal with possible business risks. The risk management department can prevent risks by discovering hidden dangers as early as possible, through sensitive detection of risk factors and careful investigation. It only looks at a kind of risk and cannot meet the requirements of risk prevention. The Basel Accord is also aware of this. It is not very useful to analyze credit risk, market risk, and operational risk alone. This requires linking multiple risks, judging their relevance, and forming a risk aggregation in order to avoid the occurrence of risks as much as possible.

After the 1990s, under the continuous promotion of financial innovation, with the rapid growth of financial derivatives and transactions, the banking industry gradually intervened in derivatives transactions. Therefore, the volatility of the financial market has an increasing impact on the banking industry. In 1997, the Southeast Asian financial crisis broke out. From the financial turmoil, it is discovered that the problems in the financial industry are not only single risks such as credit risk or market risk, but are formed by the combined effect of credit risk, market risk, and operational risk.

Bank risk aggregation refers to incorporating multiple types or sources of risk into a single metric by considering correlations between these risks (BCBS 2010). The aggregate results are crucial to calculating the accurate total economic capital (EC) against potential total losses from dependent risk types, especially proven by the subprime crisis (Caporale et al. 2017; Bongini et al. 2018). Basel II proposes that commercial banks should measure market risk, credit risk and operational risk, and implement integrated risk management for these risks (BCBS 2006). After the sub-prime crisis, the Basel Committee on Banking Supervision issued the Basel III in 2010, which significantly increased the tier one capital adequacy ratio to deal with the overall risks faced by banks (BCBS 2010).

Specifically, In the early stage, relevant scholars believed that these risks were not related to each other, so they were measured separately and managed separately, and the economic capital of various risk types was estimated separately (BCBS 2009). However, the various risks of banks are not independent of each other, and there is a complex correlation between them (Imbierowicz and Rauch 2014). Their role in fueling the flames may even trigger a financial crisis and seriously threaten the stability of the financial system. During the subprime mortgage crisis, many European and American banks went bankrupt. With the poor management of bank operators, the traditional risk management model has been unable to meet the needs of bank risk management under the new situation, focusing on single risk and ignoring the correlation of multiple risks. Bank risk aggregation has gradually become the focus of academic and industry.

Thus, to support the capital management and capital allocation decisions of top managers, it is not enough to evaluate and control different types of risks separately. Commercial banks must measure the overall risk, that is, they need to solve the problem of risk aggregation. Bank risk aggregation is a process of calculating the overall risk of a bank after considering the correlation between different risks faced by the bank. Bank risk aggregation is of great significance for maintaining the stable operation of banks and preventing the outbreak of financial crisis (Berger et al. 2016). The aggregate results are crucial to calculating the accurate total economic capital (EC) against potential total losses from dependent risk types, especially proven by the subprime crisis (Caporale et al. 2017; Bongini et al. 2018).

In addition, regulators also attach great importance to bank risk aggregation. According to the New Basel Capital Accord, commercial banks should measure market risk, credit risk and operational risk, and implement integrated risk management for these risks (BCBS 2006). After the financial crisis, the Basel Committee on banking supervision (BCBS) released the third version of Basel Accord in 2010, which greatly increased the tier 1 capital adequacy ratio to cope with the overall risks faced by banks (BCBS 2010). In the *Development of Risk Aggregation Modelling*, risk aggregation is redefined. Risk aggregation solves the problem of robustness to a certain extent. Through a large number of interviews and investigations, the report introduces the method for companies and regulatory agencies to summarize risks, and proposes what effect will be obtained by summarizing risks under this method. In addition, the improvement measures that may be achieved in various fields are also proposed to promote the risk aggregation of the company and the regulatory agency.

Thus, with the establishment of a more complete capital adequacy ratio framework in Basel II, banks' risk management capabilities will also be strengthened. Basel II has effectively promoted banks' integrated risk management capabilities, encouraged attention to the relationship between risks, and improved the level of risk assessment. In modern life, as the types of risks gradually increase, banks are exposed to more and more risks, and many risks will inevitably influence each other. To strengthen the ability to identify risks, the Basel Committee on Banking Supervision believes that capital regulations and modern risk management should be combined, and risks should be emphasized through the disclosure of related risks and capital information.

Considering risk aggregation is of great significance to the development of the banking industry. With the development of the financial industry, the banking business has increased, and the links between risks have become closer. There is a correlation between credit risk and market risk; there is a correlation between credit risk and operational risk; there is a correlation between market risk and operational risk and even between risk factors. Because banks have a wide range of risk types and complex structures, it is often not a simple matter to analyze the bank's risk integration in detail. For banks, only by measuring bank risks more accurately and fully analyzing various types of risks, can they ensure that they develop in a positive direction and reduce bank risks.

References

Allen D, Faff R (2012) The global financial crisis: some attributes and responses. Acc Finance 52(1):1–7. https://doi.org/10.1111/j.1467-629X.2011.00416.x
Basel Committee on Banking Supervision (1988) International convergence of capital measurement and capital standards. Bank for International Settlements, Basel, Switzerland
Basel Committee on Banking Supervision (2006) International convergence of capital measurement and capital standards: a revised framework. Bank for International Settlements, Basel, Switzerland
Basel Committee on Banking Supervision (2008) Principles for sound liquidity risk management and supervision. Bank for International Settlements, Basel, Switzerland
Basel Committee on Banking Supervision (2009) Findings on the interaction of market and credit risk. Bank for International Settlements, Basel, Switzerland
Basel Committee on Banking Supervision (2010) Basel III: a global regulatory framework for more resilient banks and banking systems. Bank for International Settlements, Basel, Switzerland
Beltrame F, Previtali D, Sclip A (2018) Systematic risk and banks leverage: the role of asset quality. Financ Res Lett 27:113–117. https://doi.org/10.1016/j.frl.2018.02.015
Berger AN, El GS, Guedhami O, Roman RA (2016) Internationalization and bank risk. Manage Sci 63(7):2283–2301. https://doi.org/10.2139/ssrn.2249048
Bongini P, Clemente GP, Grassi R (2018) Interconnectedness, G-SIBs and network dynamics of global banking. Financ Res Lett 27:185–192. https://doi.org/10.1016/j.frl.2018.03.002
Caporale GM, Alessi M, Di Colli S, Lopez JS (2017) Loan loss provisions and macroeconomic shocks: Some empirical evidence for Italian banks during the crisis. Financ Res Lett 25:239–243. https://doi.org/10.1016/j.frl.2017.10.031
Imbierowicz B, Rauch C (2014) The relationship between liquidity risk and credit risk in banks. J Bank Finance 40:242–256. https://doi.org/10.1016/j.jbankfin.2013.11.030
Li J, Zhu X, Wu D, Lee CF, Feng J, Shi Y (2015) On the aggregation of credit, market and operational risks. Rev Quant Financ Acc 44(1):161–189. https://doi.org/10.2139/ssrn.478381
Mustika G, Suryatinc E, Hall MJB, Simper R (2015) Did Bank Indonesia cause the credit crunch of 2006–2008? Rev Quant Financ Acc 44(2):269–298. https://doi.org/10.1007/s11156-013-0406-4

Chapter 2
Research Review of Bank Risk Aggregation

2.1 Bank Risk Correlation

The first key aspect of bank risk aggregation is to determine the correlation relationships between risks needed to be aggregated. The banks in the banking system are interrelated, and risks can be transmitted through multiple channels among different banks (Kaufman and Scott 2003). Within banks, there are various types of risks, such as credit risk, market risk, operational risk and liquidity risk (Breuer et al. 2008b; Hull 2012), there is a correlation between these different risks, and because the risks can be further resolved into risk factors or risk elements, there is also a correlation between single risks.

At present, many scholars have studied bank risk aggregation from the perspective of correlation. PéRignon and Smith (2010), Huang et al. (2012), Yun and Moon (2014), Greenwood et al. (2015), Karimalis and Nomikos (2017) and Yao et al. (2017) studied the systemic risk of the whole banking industry caused by risk contagion before banks. Rosenberg and Schuermann (2006), Drehmann et al. (2010), Li et al. (2015) integrated the total risk of a single bank by considering the correlation among various types of risks within a bank, such as credit risk, market risk and operational risk. Bocker and Kluppelberg (2008), Angela et al. (2009), Guegan and Hassani (2012) and Brechmann et al. (2014) modeled the internal risk unit correlation of operational risk, and integrated the operational risk of banks. Therefore, the first key aspect of bank risk aggregation is to identify the correlation structure between different risk types. There are various relationships in bank risk, and risk can be subdivided into different levels of risk. There are correlations among different banks, different types of risks, and different sub risks and risk elements within the same type of risk. These relationships are the objects that need risk aggregation. However, there is no article to sort out how many levels of bank risk correlation are included in bank risk aggregation.

J. Li et al., *Financial Statements-Based Bank Risk Aggregation*, Innovation in Risk Analysis, https://doi.org/10.1007/978-981-19-0408-0_2

We divide the correlation relationships of bank risks into three levels. The first level is the correlation between the total risks among banks, the second level is the correlation among different types of risks within a bank, and the third level is the internal correlation of certain types of risks.

Specifically, the correlation between total bank risks belongs to the highest level correlation. There are various connections within the banking system. The first one is the connection of direct business such as interbank borrowing and interbank deposit in the interbank market (Yao et al. 2017). When a bank defaults, the risk will quickly spread through the interbank market. The second is the association of holding common assets. When asset prices are impacted, it is assumed that banks will sell their assets at a reduced price in order to maintain their original leverage, and the risk is transmitted through the cost channel of selling at reduced prices (Greenwood et al. 2015). The third kind of association is the connection of payment and settlement business. When banks encounter liquidity crisis and fail to pay on time, risks will be transmitted through payment and settlement channels. The last one is the run caused by information asymmetry (Upper and Worms 2004). Therefore, when some banks have problems, the risk will spread to other banks through various channels, which will lead to the collapse of the whole banking system.

The correlation between different types of risk belongs to the correlation of high level. According to Basel Accord, credit risk, market risk and operational risk are the three main risks faced by banks. These risks come from all kinds of businesses of the bank. Due to the complicated relationship between the banking business, all kinds of risks are interactive and inseparable. The relationship between credit risk and market risk has been widely concerned by scholars and practitioners. For example, unexpected changes in the market value of a company will lead to market risk, which in turn will affect the probability of the company's default and generate credit risk. On the contrary, if the default rate changes unexpectedly, credit risk will occur, which will affect the market value of the company and generate market risk (Jarrow and Turnbull 2000). In addition, credit risk and market risk are also closely related to operational risk. For example, banks with higher credit risk have more operational losses. For example, in the period of declaring operational risk loss, abnormal negative returns often appear, accompanied by the increase in the volume of bank stock transactions. In the case of internal fraud, the loss of bank market value is greater than the loss of published operational risk. All these indicate that there is a close correlation between market risk and operational risk.

The internal correlation of single risk belongs to the lowest level correlation. In order to explore the internal structure of risk more accurately, some types of bank risks have been further analyzed, and there is also a correlation between the risk units after analysis. Therefore, it is necessary to model the correlation between these risk units to accurately measure this kind of risk. For example, credit risk and market risk can be further analyzed. According to the definition of credit risk and market risk, we can see that credit risk and market risk come from the fluctuation of many risk factors, such as GDP, interest rate, exchange rate and stock index, and there is a correlation between these risk factors. Among the risks that can be further analyzed, the most typical one is operational risk.

The Basel Committee divides the business activities of commercial banks into eight business lines and seven loss types so that the operational risk can be divided into risks from different business lines and different loss types. The eight business lines of commercial banks are ① corporate finance; ② trading and sales; ③ retail banking; ④ commercial banking; ⑤ payment and settlement; ⑥ agency service; ⑦ asset management; and (8) retail brokerage. In addition, the business types that cannot be classified into the above eight business lines are divided into other businesses. The seven types of loss events are ① internal fraud; ② external fraud; ③ employment system and workplace security incidents; ④ customer, product and business activity events; ⑤ destruction of physical assets; ⑥ information technology system events; ⑦ execution, delivery and process management events. The eight business lines and seven loss types divide the operational risk into 56 risk units (cells).

Due to the close relationship between the bank's business activities, there is a correlation between each risk unit. The correlation between these risk units is called loss correlation in operational risk modeling. It refers to the correlation between the annual total losses of different risk units. In addition, because the same risk unit is composed of loss frequency and loss intensity, the correlation between loss frequency of each risk unit is called frequency dependence, and there is also a correlation between loss intensity, called severity dependence. Specifically, frequency correlation refers to the correlation between the number of loss events of risk units within a certain period of time. For example, in practice, it may be observed that when the number of business line risk events of commercial banks is high, the number of risk events of corporate financial business lines is also relatively high, on the contrary, when the number of commercial bank events is small, the number of corporate financial events is also small. Intensity correlation refers to the correlation between the loss size caused by a single loss event among different risk units.

In short, there are many levels of correlation between different bank risks. Through systematic induction and carding, we divide the bank risk correlations into three levels. The division of three levels of bank risk correlation makes the bank risk aggregation present the idea of progressive and gradual aggregation. From bottom to top, they are the correlation within a certain type of risk, the correlation among different types of risk within a bank, and the correlation between the risks of various banks in the banking system. Through the reasonable modeling of the internal correlation of risks, we can get the risk value of a single risk type, and then through the correlation modeling between different types of risks, and we can get the total risk value of the bank. Finally, we can get the systemic risk of the whole banking industry by modeling the correlation of the total risks of different banks.

2.2 Typical Characteristics of Bank Risk Correlation

After determining the correlation relationship between different bank risks, the second key step of bank risk aggregation is to identify the typical characteristics of correlation between risks. Whether it is the correlation between bank risks, or the correlation between risk factors and risk elements within a single risk, they all have some common, typical characteristics. In the existing literature, many scholars have found the typical characteristics of the correlation between bank risks.

Dionisio et al. (2006) and Mittnik et al. (2013) found that the relationship between bank risks is not a simple linear correlation, but a complex nonlinear correlation. Cherubini et al. (2004) found that there was a lower tail correlation between bank risks, and Hartmann (2010) found that there was an upper tail correlation between bank risks. However, the upper tail correlation and the lower tail correlation are usually asymmetric (BCBS 2010b). Li et al. (2015) found that the correlation between various types of banks had their own characteristics. Guegan and Zhang (2010) and Karimali and Nomikos (2017) found that the correlation between bank risks was not constant, and the correlation degree and structure would change over time. Therefore, because the correlation between bank risks presents various complex characteristics, such as non-linear, tail correlation, tail asymmetry, structural asymmetry, and time-varying, the second key aspect of risk aggregation is to clarify the typical characteristics of risk correlation. However, the existing literature only mentioned one or several typical characteristics. There is no summary of the typical characteristics of the correlation between risks that need to be considered in bank risk aggregation. Based on the existing research, we summarize the seven typical bank risk correlation characteristics: nonlinear, tail correlation, tail asymmetry, structural asymmetry, high dimensionality, structural diversity, and time-varying.

(1) Non-linear Characteristics

The first typical characteristic of bank risk correlation is nonlinearity. The opposite of nonlinear correlation is a linear correlation, which means that when one variable changes, another variable always changes at a fixed proportion. Since the definition of linear correlation is relatively simple and easy to identify, when it is necessary to describe the correlation between variables, the linear correlation is usually first thought of (Dionisio et al. 2006; Mittnik et al. 2013). However, in complex practical situations, linear correlation is an over simplified correlation (Moddemeijer 1999). In particular, the business involved by banks is becoming more and more complex, and there are intertwined relationships among the businesses involved in various risks. As a result, the correlation between bank risks is usually not a simple linear correlation, but a complex nonlinear relationship. Some particularly complex nonlinear relationships can not even be expressed by mathematical formulas.

(2) Tail Correlation

The second typical characteristic of bank risk correlation is tail correlation. The right part of the probability density distribution of risk is called the upper tail, and the part to the left is called the lower tail. When one risk suddenly increases, the other risk will increase significantly, which is called upper tail correlation. When one risk suddenly decreases, the other risk also decreases significantly, which is called lower tail correlation (Cherubini et al. 2004). In practice, it is often observed that under financial crisis or other pressure situations, different types of risks often increase significantly at the same time (Hartmann 2010), which indicates that there is an upper tail correlation between risks. The calculation of value at risk (VaR), which has become a standard measure of financial risk, is closely related to the upper tail of risk distribution (Hull 2012). Therefore, it is more and more important to estimate the tail of risk distribution accurately. Only by describing tail correlation accurately can VaR be calculated accurately, so tail correlation is a very important characteristic of bank risk.

(3) Tail Asymmetry

The third characteristic of bank risk correlation is tail correlation asymmetry; that is, the correlation structure of upper tail is different from that of lower tail. In practice, it is often observed that different types of risk will increase significantly under pressure, but it is rarely observed that different types of risks will significantly reduce at the same time, which indicates that in general, the correlation between the upper tail and the tail is stronger than that of the lower tail (BCBS 2010b). In addition, even if the upper tail correlation and the lower tail correlation exist simultaneously, there will be differences between their forms and sizes, which is generally impossible to be completely consistent. Therefore, the upper tail correlation and the lower tail correlation are usually asymmetric (Li et al. 2015). This feature and the second feature are the characteristics of the tail structure of bank risk. Whether the tail structure of risk correlation can be well described plays an important role in accurately measuring risk.

(4) Structural Asymmetry

The fourth important characteristic of bank risk correlation is structural asymmetry. Banks usually face two or more risks at the same time, such as credit risk, market risk and operational risk. These risks are correlated with each other. For example, there are correlations between credit risk and market risk, between credit risk and operational risk, and between market risk and operational risk (BCBS 2003; Li et al. 2015). Because of the different meanings and sources of various risks, the correlations between these risks have their own characteristics and are different from each other. Therefore, we can not use the same correlation structure to describe the correlation between two risks, which is the structural asymmetry of risk correlation. Within the same type of risk, there will be approximately structurally symmetric risks under special circumstances. For example, the interaction between homogeneous risks in credit portfolios is the same (BCBS 2010b). However, the interaction between different types of risk is generally different, and banks need to consider the structural asymmetry when modeling the correlation of different types of risk.

(5) High Dimensionality

The fifth characteristic of bank risk correlation is the high dimension. There are many objects of bank risk aggregation. When studying the systemic risk of the banking industry, it is necessary to consider the correlation among multiple banks in the banking system (Yao et al. 2017). As for the internal risks of banks, in addition to the three main types of risks, credit risk, market risk and operational risk, banks are also faced with liquidity risk, country risk, reputation risk, legal risk and strategic risk (BCBS 2006). In addition, operational risk is divided into 56 related risk units (cells) by eight business lines and seven loss types. The interaction between these 56 risk units should be considered when integrating operational risk. Therefore, banks need to consider the high dimension of risk objects in risk aggregation.

(6) Structural Mixture

The sixth important characteristic of bank risk correlation is structural mixture. For the overall data, different parts of risk data may show different correlation structures. Therefore, the overall data presents a mixture of multiple related structures. For example, for some risk data, the correlation structure of some data is A, the correlation structure of some data is B, and the correlation structure of some data is C. Because the correlation structures of different parts of the data are different, the overall data hybridized a variety of related structures, reflecting the characteristics of the diversity of related structures. Therefore, it is necessary to consider the correlation structure of different parts of data in the process of bank risk aggregation (Zhu et al. 2018).

(7) Time Varying

The above six typical characteristics of bank risk are static, while the seventh typical characteristic of bank risk correlation is dynamic: time-varying. In financial risk management, the correlation between risks is not constant. With the emergence of market shock force, the adjustment of national macroeconomic policies, the change of bank market rules, and the emergence of political events may affect the correlation structure and degree between the financial market and financial asset portfolio (Bauwens et al. 2006). Generally, this effect can be divided into two situations: one is that although the correlation structure has not changed, the degree of correlation changes with time. Secondly, under the stable and extreme conditions of the market, the correlation structure between bank risks may change significantly, and the uncorrelated risks may become correlated in extreme cases (Andrievskaya and Penikas 2012). Therefore, the correlation between bank risks will show dynamic changes, so we should capture the dynamic correlation structure and correlation degree between risks in the process of bank risk aggregation.

2.3 Bank Risk Aggregation Approaches

As one of the most important financial institutions in the economic system, banks play an important role in realizing the rational allocation of financial resources and promoting the rapid development of the real economy (Allen and Faff 2012). However, the business activities of banks inevitably produce various types of risks, such as credit risk, market risk and operational risk (Basel Committee on Banking Supervision, BCBS for short 2006). At first, these risks were considered uncorrelated, so they were measured separately and managed separately, and the economic capital of various types of risks was estimated separately (BCBS 2009). However, all kinds of risks of banks are not independent, and there are complex correlation relationships between them, which will transform and influence each other, making the original risks tend to enlarge or decrease (Malevergne and Sornette 2006; Breuer et al. 2008a; Imbierowicz and Rauch 2014). These risks may cause the bank to suffer heavy losses and even face the risk of bankruptcy. They may even lead to financial crisis and threaten the stability of the financial system. The subprime crisis, which started in the United States in 2007, highlights the importance of risk aggregation methods considering the correlation between risks for risk management and crisis prevention (Berger et al. 2016).

The accuracy of the risk aggregation results is determined by the ability of the bank risk aggregation method to describe the complex characteristics of the correlation relationship. Only by describing these complex characteristics can we integrate bank risk accurately. Therefore, the third key aspect of bank risk aggregation is whether the bank risk aggregation method can comprehensively describe the typical characteristics of risk correlation.

A lot of researches on bank risk aggregation are carried out on the basis of describing the correlation between risks. However, the existing risk aggregation methods have different ability to describe these typical characteristics. Specifically, the widely used simple addition method, fixed coefficient method and variance/covariance method cannot fully capture the nonlinear characteristics between risks, but mainly describe the linear relationship between risks (Li et al. 2012). Recently, Copula function group has become the mainstream method of bank risk aggregation, including a variety of copula, which can describe a variety of different correlations. Rosenberg and Schuermann (2006) and Inanoglu and Jacobs (2009) used Gaussian copula to describe the nonlinearity and structural asymmetry between risks. The t-copula used by Morone et al. (2007) and Tang and Valdez (2009) can not only describe the nonlinearity and structural asymmetry, but also describe the tail correlation between risks.

Li et al. (2015) used Gumbel copula and Clayton copula to characterize risk-related tail asymmetry. When there are too many risk aggregation objects, Guegan and Hassani (2012) and Brechmann et al. (2013) used Vine copula to solve the high-dimensional problem caused by risk aggregation objects. In order to capture the dynamic changes of the correlation between risks, Pérignon and Smith (2010), Huang et al. (2012), Karimalis and Nomikos (2017) used time-varying copula method, Guegan and Zhang (2010), Andrievskaya and Penikas (2012) variable structure copula. However, few papers summarize the ability of risk aggregation methods to describe the typical characteristics of complex correlation.

With the gradual deepening of the understanding of risk correlation characteristics, the ability of the risk aggregation method to describe the typical characteristics of correlation is also gradually strengthened. In order to reflect the evolution of the risk aggregation method, we divide the bank risk aggregation method into three times according to the ability to describe the typical characteristics of bank risk correlation: the correlation coefficient method in the 1.0 era, which mainly describes the linear correlation risk aggregation, and the basic copula class in the 2.0 era which can depict non-linear, tail related, tail asymmetric and structural asymmetry 3.0 complex copula methods, which can depict complex high-dimensional, structural diversity and time-varying.

The ability of bank risk aggregation methods in these three risk aggregation aspects to describe risk correlation characteristics is shown in Table 2.1. As shown in Table 2.1, the typical static characteristics of risk correlation include nonlinearity, tail correlation and tail asymmetry in tail characteristics, structural asymmetry and high dimensionality. Nonlinear correlation can be refined into positive nonlinear correlation and negative nonlinear correlation, and tail correlation can also be refined into positive tail correlation and negative tail correlation. The dynamic characteristics are time-varying. The symbol "√" indicates that the method can describe the corresponding characteristics, and the symbol "×" indicates that the method cannot depict the corresponding characteristics.

2.3.1 *Risk Aggregation 1.0 Era—Correlation Coefficient Method*

In the era of risk aggregation 1.0, the risk aggregation method is a kind of correlation coefficient method, which mainly describes the linear correlation between risks, and does not fully capture the nonlinear correlation between risks. The representative methods are the simple addition method, fixed effect coefficient method, variance skew difference method and mutual information entropy method.

The simple addition method is the earliest method, which assumes that there is a complete linear positive correlation between risks, and does not specifically consider the correlation characteristics between risks (Li et al. 2012). It can not describe the nonlinear correlation and tail correlation and the asymmetric tail correlation. At the same time, because the correlation coefficient of all kinds of risks is set as 1, it can not describe the asymmetric correlation between different types of risks. The simple addition method does not consider the more complex correlation characteristics such as high dimensionality and time-varying. In short, the simple addition method can not describe the seven typical characteristics of bank risk, so in Table 2.1, the whole column corresponding to this method is "×".

Since the simple summation approach assumes that there is a complete positive correlation between risks, that is, the marginal risks are directly added to get the total risk. Therefore, it is generally regarded as a conservative method (Embrechts et al. 2003) and ignores the decentralization effect. However, a certain amount of literatures show that there is a large diversification effect between different risks (Rosenberg and Schuermann 2006; Drehmann et al. 2010). Therefore, the later research improved this method, that is, subtracting a certain proportion from the simple addition results as the result of the decentralization effect. This improved method is often referred to as the "fixed coefficient method" (Li et al. 2012). The biggest problem of this method is to determine a suitable proportion of decentralization effect, but there is no widely accepted method to determine this proportion. Like the simple addition method, the fixed coefficient method can not describe the seven typical characteristics of bank risk. Therefore, in Table 2.1, the whole column corresponding to this method is "× ".

Table 2.1 The ability of methods to capture the complex characteristics of bank risk correlation

Main models				Correlation coefficient type				Basic copula type					Complex copula type			
Characteristics of correlation				Simple summation approach	Fixed effect coefficient approach	Variance/covariance approach	Mutual information entropy approach	Elliptical copula		Archimedean copula			Vine copula	Hierarchical copula	Mixed copula	Time-varing copula
								Gaussian	t	Gumbel	Clayton	Frank				
Static characteristics	Non-linear	Positive correlation		×	×	×	√	√	√	√	√	√	√	√	√	√
		Negative correlation					×			×	×		√	√	√	√
	Tail characteristics	Tail correlation	Upper tail correlation	×	×	×	×	×	√	√	x	×	√	√	√	√
			Lower tail correlation							×	√		√	√	√	√
		Tail asymmetry							×	√	√		√	√	√	√
	Structural asymmetry			×	×	√	√	√	√	×	×	×	√	√	√	√
	High dimensionality			×	×	×	×	×	×	×	×	×	√	√	×	×
	Structural diversity			×	×	×	×	×	×	×	×	×	×	×	√	×
Dynamic characteristics	Time-varying			×	×	×	×	×	×	×	×	×	×	×	×	√

Note The symbol "√" indicates that the method can describe the characteristic, and the symbol "×" indicates that the character cannot be described

Variance/covariance method was once the most widely used method. According to the survey conducted by the international financial risk Institute Foundation and chief risk officer Forum (IFRI/CRO) in 2007, 75% of banks used variance/covariance method to describe bank risk correlation (IFRI/CRO 2007). This method uses a linear correlation coefficient to describe the correlation between risks, which is a step forward than the simple addition method, and considers the linear correlation between risks (BCBS 2010b). This method also uses a linear correlation coefficient to describe the correlation between risks, so it can not describe nonlinear correlation, tail correlation and asymmetric tail correlation. However, compared with the simple addition method, this method uses the actual estimated linear correlation coefficient, which allows the correlation coefficient between risks to be inconsistent. Therefore, this method can describe the different correlations between risks, that is, to satisfy the structural asymmetry. Thus, in Table 2.1, the penultimate line corresponding to this method is "√", and the other corresponding lines are "×".

The mutual information entropy method calculates the mutual information between two variables, and then becomes the global correlation coefficient between the two variables after standardization. Because mutual information contains all the relevant information between two variables, including linear correlation and nonlinear positive correlation (Dionisio et al. 2006), the global correlation coefficient is considered to be able to capture the global correlation between the two variables, and is considered as an extension of the concept of linear correlation in the field of nonlinearity (Moddemeijer 1999). Compared with variance/covariance method, which only considers linear correlation and imposes the assumption that marginal risk distribution obeys normal distribution, the global correlation coefficient can capture the global correlation between two variables, and its calculation process does not require any theoretical probability distribution or model assumption (Darbellay 1998).

Therefore, in the framework of variance/covariance, the global correlation coefficient can be used to replace the linear correlation coefficient, so as to maintain the simplicity of the variance/covariance method and expand the ability of the method to describe nonlinear correlation. Similar to variance/covariance, the global correlation coefficient of the mutual information entropy method allows the correlation coefficients between risks to be inconsistent. Therefore, this method can describe different correlation relationships between risks, that is, satisfying structural asymmetry. Therefore, in Table 2.1, the corresponding nonlinear positive correlation and structural asymmetry are "√", and the other corresponding rows are "×".

In conclusion, the correlation coefficient risk aggregation method in the era of risk aggregation 1.0 only considers the linear correlation and nonlinear positive correlation between risks, and can not fully capture the nonlinear correlation between risks. Therefore, the correlation coefficient method is generally considered to underestimate the diversification effect between risks (Embrechts et al. 1999; Rachev et al. 2005).

2.3.2 *Risk Aggregation 2.0 Era—Basic Copula Approach*

In the era of risk aggregation 2.0, the basic copula method (Rosenberg and Schuermann 2006) was introduced. Copula function was used to capture the correlation between risks, and nonlinear, tail characteristics and structural asymmetry were considered. Compared with the linear correlation coefficient method in the era of risk aggregation 1.0, the Copula function can capture the complete correlation structure between random variables (Ward and Lee 2002; Weiß 2013), including linear correlation and nonlinear correlation. In addition, Copula functions have a large family of functions. Various copula functions have their own characteristics and can describe a variety of different correlation relationships.

Sklar (1959) defined copula model, Nelsen (1999) systematically summarized the relevant theory of copula model, Embrechts et al. (1999) first applied this method to the financial field. Since then, copula method has been widely used in the field of bank risk aggregation. Copula method can be applied to integrate different types of risk, especially credit risk, market risk and operational risk. They include Dimakos and Aas (2004), Bouye et al. (2002), Rosenberg and Schuermann (2006), Inaoglu and Jacobs (2009), Li et al. (2015). Dimakos and Aas (2004) compared the copula method with the simple addition method, and found that the simple addition overestimated the total risk of banks because it did not consider the risk dispersion effect caused by the correlation between risks. There are also studies focusing on how to use a variety of different types of Copula Functions to characterize the internal correlation of operational risk. For example, Bocker and Kluppelberg (2008) used levy copula to describe the frequency correlation and intensity correlation between different risk units. Angela et al. (2009) used t-copula to describe the correlation between different loss types of operational risk.

Figure 2.1 shows the copula functions commonly used in the characterization of bank risk correlation, including Gaussian copula and t-copula in elliptical copula and

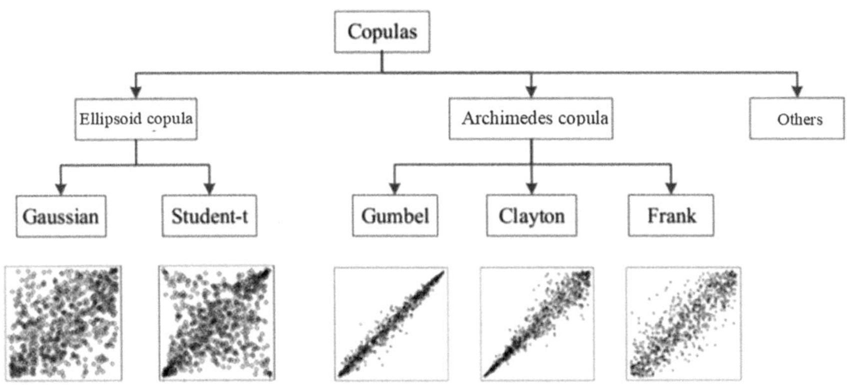

Fig. 2.1 Copula function commonly used in bank risk correlation modeling

Gumbel copula, Clayton copula and frank copula in Archimedean copula (Cherubini et al. 2004; Corrigan et al. 2009).

In the elliptical copula, Gaussian copula can describe the nonlinear correlation, including positive correlation and negative correlation, allow the inconsistent correlation between marginal risk distributions, and describe the structural asymmetric correlation between multiple risks. However, the copula can not capture tail correlation, and the structure of upper tail and lower tail is the same. Therefore, the main defect of the Gaussian copula is that it can not describe the tail characteristics between risk correlations (Nelsen 1999). As a typical elliptical copula, t-copula can describe nonlinear correlation and structural asymmetry. Compared with Gaussian copula, t-copula can also capture tail correlation, including positive correlation and negative correlation. Therefore, among the commonly used copula, t-copula can describe the most relevant features, and it is also the most widely used copula in the existing literature (Rosenberg and Schuermann 2006; Morone et al. 2007; Tang and Valdez 2009; BCBS 2010b). However, the upper and lower tails of the t-copula are symmetric, so it can not describe asymmetric tail dependence (Cherubini et al. 2004). Gumbel copula, Clayton copula and frank copula are three typical Archimedean copulas. One common feature of Archimedean copulas is that they have only one parameter, which can only describe the correlation between homogeneous risks, so they can not satisfy structural asymmetry (BCBS 2010b).

Gumbel copula can describe the nonlinear correlation and tail correlation, and Gumbel copula can sensitively capture the change of upper tail correlation, but difficult to capture the change of lower tail correlation so that it can reflect the tail asymmetry. But it can only capture the positive nonlinearity and tail correlation, and cannot describe the negative correlation. Similar to the Gumbel copula, the Clayton copula can also capture nonlinear correlation and tail correlation, but cannot describe negative correlation. The difference is that Gumbel copula can capture upper tail correlation, while Clayton is on the contrary, it is very sensitive to lower tail correlation. Frank copula can describe nonlinear correlation, including positive and negative correlations, but can not capture tail characteristics.

In short, compared with the correlation coefficient method in the era of risk 1.0, although the basic copula method in the era of risk aggregation 2.0 has begun to consider the typical characteristics of nonlinear correlation, tail correlation, tail asymmetry and structural asymmetry, high dimensionality, structural diversity and time-varying can not be described. Therefore, there is no basic copula function that can fully describe the typical characteristics of bank risk correlation. In Table 2.1, there is no method corresponding to all columns are "√".

2.3.3 Risk Aggregation 3.0 Era—Complex Copula Approach

In the era of risk aggregation 2.0, various basic copula functions have been close to the requirements of comprehensively describing the characteristics of risk correlation. In the era of risk aggregation 3.0, all kinds of complex copula, including hierarchical copula, Vine copula, hybrid copula, time-varying copula and variable structure copula, have been introduced into the field of bank risk correlation modeling to consider more complex risk correlation characteristics: high dimension, structural diversity and timely variability.

(1) Consideration of High Dimensionality

For the high-dimensional characteristics of bank risk dependencies, although the theory of multivariate copula is feasible, because of the complex correlation among various risk types, it is difficult to directly use a multivariate copula function for correlation modeling, whether it is parameter estimation or model selection. The accuracy and flexibility of multivariate Copula Functions in describing multivariate joint distribution have various limitations (McNeil et al. 2005).

Therefore, in view of the high-dimensional characteristics of bank risk dependencies, hierarchical copula and vine copula are introduced respectively to solve the high-dimensional risk aggregation problem. The hierarchical copula is to layer the risk types that need to be integrated, and then achieve the purpose of dimension reduction through layer-by-layer aggregation. The Vine copula uses the Vine structure to reduce the high-dimensional copula function to a series of two-dimensional Copula Functions. The following two kinds of complex copula are introduced in detail.

Hierarchical copula intuitively describes the hierarchical aggregation among risks by integrating an "aggregation tree". By analyzing the correlation between risks, the risks are grouped and placed at different levels. Through hierarchical aggregation, the dimensions of the related structures which describe all risks are reduced to depict the relationship among risks layer by layer to achieve the purpose of reducing the dimensions of risk aggregation objects. The structure of hierarchical copula function is more flexible, allowing different types of Copula to be used in risk aggregation of different layers, which makes the selection of Copula more diverse and flexible, and the parameter estimation is more simple, which can more accurately capture the correlation between various risk types.

The key of the hierarchical copula method is to determine the aggregation "tree", that is, how to carry out hierarchical aggregation. Bürgi et al. (2008), Bruneton (2011) and Arbenz et al. (2012) gave different methods to determine the ensemble tree. In terms of application, Abadi (2015) gives the specific steps of how to use the hierarchical copula method proposed by Arbenz et al. (2012). Gaisser et al. (2011) integrated the risk of German banks with the hierarchical copula.

Another way of dimensionality reduction is to separate the correlation structure between multiple risks into two related structures through a "Vine" structure, so as to reduce the dimension of multivariate copula into the product of a series of two-dimensional copula. Therefore, only a series of binary copula functions need to be estimated separately, which greatly reduces the difficulty of model estimation. Moreover, the structure of the Vine copula function is more flexible. The correlation between risks can be better fitted by selecting different Vine decomposition structures and different binary copula functions.

Vine copula was proposed by Joe (1996) and Bedford and Cooke (2002). Teng's ability to model related complex structures has been supported by many empirical studies, and has been increasingly used in the calculation of high-dimensional copula models in recent years (Aas et al. 2009; Guegan and Maugis 2009; Brechmann et al. 2012). In the field of banking risk aggregation, Hofmann and Czado (2010) used the Vine copula to study the risk management of the portfolio, and compared with the traditional multi copula. The empirical results show that because Vine copula is more flexible in constructing multivariate correlation structure, it can depict the symmetric or asymmetric correlation structure between two variables and whether it has tail distribution, which affects the accuracy of VaR calculation. Brechmann et al. (2013) and Pourkhanali et al. (2016) modeled the risk correlation among financial institutions through Vine copula, and then measured the systemic risk of global banks and insurance companies. Guegan and Hassani (2012) and Brechmann et al. (2014) used the Vine copula to model the internal correlation of operational risk. By comparing the characteristics of Archimedean copula, Gaussian copula, t-copula and Vine copula, Brechmann et al. (2014) found that Vine copula can capture the correlation between operational risk business lines more flexibly.

In conclusion, hierarchical copula and Vine copula, which can describe the high-dimensional characteristics of risk correlation, can describe the non-linearity, tail correlation, tail asymmetry, structural asymmetry and high-dimensional between risks. As reflected in Table 2.1, except for the structural diversity and time varying of the last two rows of Copula methods, the other corresponding rows are "√".

(2) Consideration of Structural Mixture

In the overall data, the correlation structure of different parts of the data may be different, and the overall data presents a mixture of multiple related structures. Some are tail and some data are asymmetric. Therefore, in order to accurately describe the risk correlation of the overall data, it may be necessary to capture the typical characteristics of bank risk correlation, such as non-linearity, tail correlation, tail asymmetry and structural asymmetry, but the existing basic class copula cannot fully capture these typical characteristics of bank risk correlation. Therefore, a single copula function cannot describe the relevant structure of the overall data.

Nelsen (1999) proved that the mixture of Copula is still copula, and named it mixed copula (mixture copula or mixed copula). Mixed copula can produce related structures that do not belong to any known single copula (Chen and Tu 2013). Through reasonable combination of the copula, its greatest advantage is that it can nest a variety of different correlation structures at the same time, resulting in the typical characteristics with enough flexibility to comprehensively describe the correlation (Hu 2006; Rodriguez 2007; Li et al. 2014a, 2014b; Huang and Wu, 2015).

According to the principle that mixed copula has the characteristics of the copula, it can comprehensively describe the typical characteristics of non-linearity, tail correlation, tail asymmetry and structural asymmetry between bank risks, and can fully capture the typical characteristics of bank risk correlation, so as to realize the accurate characterization of bank risk correlation. By mixing copulas with different advantages as much as possible, we can characterize all the static characteristics and get a list of mixed copulas with all "√" except the characteristics of "time varying" and "high dimensionality" in Table 2.1. Zhu et al. (2018) mixed Gaussian copula, Gumbel copula and Clayton copula. After mixing, Gaussian Gumbel Clayton copula can capture both upper tail correlation and lower tail correlation, and allow tail asymmetry. This paper also uses the mixed copula to describe the correlation among credit risk, market risk and operational risk of China's listed banks, and carries out risk aggregation.

(3) Consideration of Time Varying

In the era of risk aggregation 2.0, the Copula function has fixed parameters, which cannot well describe the time-varying of risk correlation (Busetti and Harvey 2010). Therefore, in the era of risk aggregation 3.0, the description of risk correlation has expanded from static to dynamic. At present, there are two main types of dynamic copula models: one is the time-varying copula model; the other is variable structure copula model. The function type of time-varying copula model does not change. The change of correlation among variables is determined by defining the evolution equation of related parameters with time. The variable structure copula model divides different wave regions or wave states by some specific theoretical methods, and then describes different wave levels with different Copula Functions (different forms of functions or different parameters). The dynamic change of the correlation between the following variables.

When using the time-varying copula model (TVC) to model correlation, the focus is to study the time-varying characteristics of Copula function parameters. Patton (2001), Engle (2002), Hafner and Reznikova (2010) and Hafner and Maner (2012) gave the evolution equations of parameters related to different Copula Functions. Engle (2002) and Patton (2006a, 2006b) used seven different time-varying copula models to estimate the dynamic correlation structure. Hafner and Reznikova (2010) and Hafner and Maner collected various forms of Copula and time-varying structure, and compared the applicability of eight different time-varying copula models.

In the study of the time-varying copula, Pérignon and Smith (2010) used dynamic conditional correlation (DCC) copula to integrate the risks of American commercial banks, and found that the average risk dispersion effect was 40%. Huang et al. (2012) and Yun and Moon (2014) used DCC to measure the systemic risk of banking industry in Asia and the Pacific region and South Korea, respectively. Karimalis and Nomikos (2017) used the time-varying copula method to measure the systemic risk of European banks.

Guegan and Zhang (2010) proposed a variable structure copula method which can consider both parameter changes and copula type changes, which can accurately describe the dynamic correlation between data. Since there are different correlation structures in different states, Pelletier (2006) first introduced mechanism transition into correlation. Garcia and Tsafack (2008) and Chollete et al. (2009) established the copula model of mechanism transformation. In the research of bank risk aggregation using variable structure copula, Andrievskaya and Penikas (2012) used variable structure copula to describe the dynamic correlation of marginal risk, integrated the systemic risk of the Russian banking system, and found that the correlation structures between risks are related to the situation of the banking industry, and there are obvious differences between stable state and extreme state.

In short, time-varying copula and variable structure copula extend the description of risk correlation from static to dynamic, and can describe the time-varying characteristics of risk correlation. At the same time, by expanding different basic Copula Functions, we can also characterize the non-linearity, tail dependence, tail asymmetry and structural asymmetry among risks. However, time-varying copula and variable structure copula do not consider the high-dimensional and structural diversity between risks. Therefore, in the table, except for the second and the third line from the bottom are "×", the other corresponding rows are "√".

To sum up, in theory, the mixed Vine copula can best describe a variety of complex static characteristics of bank risk correlation, the high-dimensional characteristics of risk aggregation object and the high-dimensional characteristics of related risk structure can be described. It is suggested that the mixed Vine copula be dynamic, and further depict the time-varying characteristics of risk correlation, so as to fully capture the seven characteristics of risk correlation Sex: nonlinearity, tail correlation, tail asymmetry, structural asymmetry, high dimension, structural diversity and time-varying.

On the whole, based on the existing research, we divide the bank risk aggregation into three key aspects, namely, the determination of bank risk correlations, the identification of typical correlation characteristics, and the ability of risk aggregation methods to describe the typical characteristics of correlation. These three key aspects determine the accuracy of bank risk aggregation. The determination of bank risk correlations is the basis of studying bank risk aggregation, and identifying the correlation characteristics between bank risks is the basis for further characterization. The ability of the risk aggregation method to describe typical characteristics of correlation determines whether the final aggregation result is accurate. However, the existing literature has not systematically summarized the three key aspects of bank risk aggregation.

Therefore, in view of the three key aspects of bank risk aggregation, we summarize the related research on bank risk aggregation under correlation from three levels: bank risk correlation, typical characteristics of correlation and risk aggregation method. The bank risk correlations of different levels and sorts out various complex characteristics of bank risk correlation are summarized systematically.

2.4 Risk Data in Bank Risk Aggregation

Risk data is an important basis for accurately measuring the overall risk of banks. Lack of data will lead to many risk aggregation methods that can not get accurate results of bank risk aggregation. A major obstacle to bank risk management is the problem of data availability (BCBS 2010a; Galloppo and Previati 2014). Banks face different types of risks, such as credit risk, market risk and operational risk (Brockmann and

Kalkbrener 2010). Sound risk management of banks is based on comprehensive measurement, management and analysis of these risks (Grundke 2013). However, many quantitative approaches typically fail due to the poor quality and low quantity of data available in the banks (Bignozzi and Tsanakas 2014; Aas et al. 2007). An obvious example is that notable data deficiencies resulted in the poor performance of risk management methods in the subprime crisis of 2007 (BCBS 2010a).

Therefore, in order to solve the problem of data missing in bank risk aggregation, scholars collect risk data from different ways. At present, the data applied in the research of bank risk aggregation measurement are mainly divided into the following five categories.

2.4.1 Internal Loss Data

In order to meet the regulatory requirements of the Basel Capital Accord, banks will accumulate risk data internally, and financial regulators will directly collect risk data from banks for supervision. For example, for the operational risk of data loss problem, some banking union databases have been formed, such as ORX (Operational Riskdata Exchange) database, DIPO database (Database Italiano Perdit Operational), DSGV database (Deutsche Sparkassen und Giroverb), GOLD database (Global Operational Loss Database), etc. Different banks form alliances and share internal operational risk loss data with each other. Among them, the ORX database, established in 2002, is the largest and most mature operational risk bank alliance database in the world. Its members include 87 banks from 23 countries. As of June 30, 2016, the ORX database included 525,395 operational risk loss events (Wei et al. 2018).

In 2001, 2002 and 2008, Basle Committee on banking supervision carried out three worldwide collections of operational risk data (BCBS 2003, 2009) (Fig. 2.2). In 2008, 121 banks from 17 countries submitted internal loss data, totaling 10.6

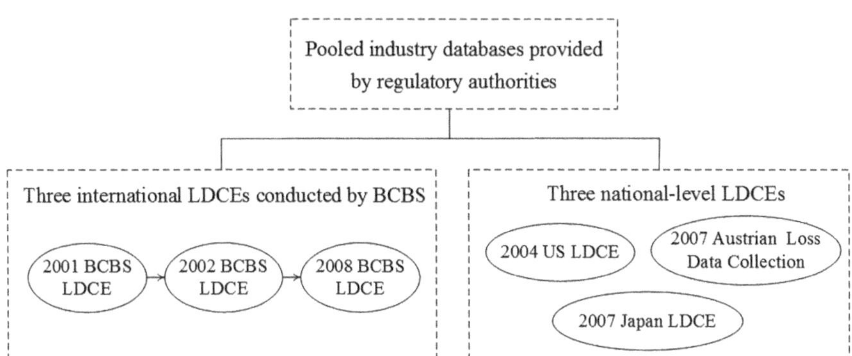

Fig. 2.2 Six pooled industry databases from BCBS and national regulatory authorities of US, Japan and Austria

million operational risk loss events. The U.S. regulatory authorities and Japanese regulatory authorities also conducted operational risk data collection in 2004 and 2007, respectively (EL Gamal et al. 2007). Participation in the BCBS LDCEs was voluntary and banks were asked to categorize individual loss events according to Basel II-defined eight business lines (BLs) and seven event types (ETs).

However, it is noteworthy that the LDCEs conducted by regulatory authorities are still very few. Regulatory authorities of most countries, especially developing countries have not begun to collect operational loss data yet. Besides, these data are generally difficult to obtain by scholars, and are only used by Union member banks and regulatory authorities.

Besides, The operational loss database provided by a consortium of financial institutions is comprised of internal loss data of member financial institutions joining the consortium. Members are allowed to use consortium data to supplement their own internal data (Figini et al. 2010). Since data standardization can enforce the completeness, accuracy, and consistency of data (Yang et al. 2014), so member financial institutions are required to submit data formatted to meet a common reporting standard (Kiss and Homolya 2014).

As shown in Fig. 2.3, we found that ten consortium databases have been constructed up to now. Eight of them are banking consortium databases, including ORX-GBD (Operational Riskdata eXchange Global Banking Database), DIPO (Database Italiano delle Perdite Operative) database, DakOR (Daten Konsortium operationelle Risiken) database, DSGV (Deutsche Sparkassen-und Giroverband) database, Global Operational Loss Database (GOLD), Operational Loss Data Sharing

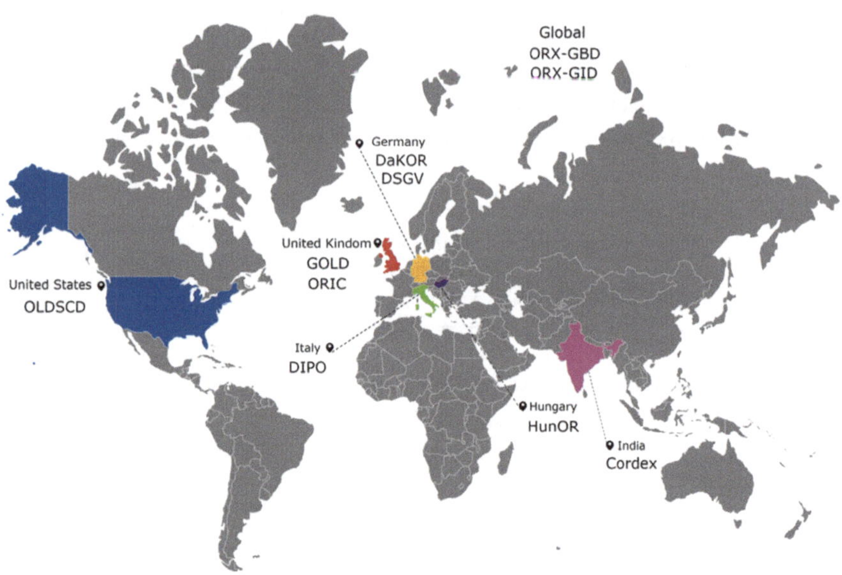

Fig. 2.3 A world distribution map of 10 consortium databases summarized in this paper

Consortium Database (OLDSCD), Cordex (Credit Operational Risk Data Exchange) database and HunOR (Hungarian Operational Risk) database. The other two are insurance consortium databases, ORX-GID (ORX Global Insurance Database) and ORIC (Operational Risk Insurance Consortium) database.

2.4.2 *Simulated Data*

Some scholars use the factor model to generate rich risk data for bank risk aggregation (Dimakos and Aas 2004; Aas et al. 2007). The fluctuation of risk factors will lead to the loss of banks. For example, market risk factors include equity price, interest rate, exchange rate and its volatility. Credit risk is related to credit factors such as credit spread and default rate (Grundke 2010; Baselga et al. 2015). After establishing the relationship between risk factors and risk profit and loss through regression technology, the profit and loss of various risks can be simulated, and abundant simulated risk data can be obtained (Alexander and Pezier 2003; Rosenberg and Schuermann 2006). However, this kind of simulation data is difficult to truly measure the bank risk (Li et al. 2018).

2.4.3 *Risk Statistics Data from Existing Literature*

Some studies directly obtain risk statistics data (i.e. mean, variance, median and mode) from academic researches, expert judgments and surveys. By reviewing a lot of relevant papers, correlations between different risk types and daily probability and loss of operational risk occurrence can be found (Rosenberg and Schuermann 2006; Li et al. 2015). Besides, experienced experts' knowledge is also heavily engaged in the risk management process (BCBS 2010a). For example, risk managers at DnB NOR felt that they had a relatively clear opinion on the size of the most frequent operational risk loss, i.e. the mode of the loss distribution (Aas et al. 2007). Third, surveys conducted by supervisory authorities, such as BCBS and International Financial Risk Institute and the Chief Risk Officers Forum (IFRI/CRO), provide valuable reports on bank risk.

For example, Elsinger and Lehar (2006) gave the average and standard deviation of credit, market and operational risks. The shapes of these three risk distributions are also discussed in the literature. Generally, we assume that credit risk follows beta distribution (Dimakos and Aas 2004; Aas et al. 2007), market risk follows a normal distribution (Rosenberg and Schuermann 2006; Hull 2012), and operational risk follows lognormal distribution (Aas et al. 2007; BCBS 2011). Based on these common distribution assumptions and collected risk statistics, the key parameters of risk distribution can be obtained through mathematical derivation, and then the credit, market and operational risk distribution can be determined. Although the risk distribution can be obtained based on a small number of risk statistics, it is too dependent on the assumption of risk distribution. If the assumption of risk distribution is not correct, then the risk distribution obtained will deviate from the real distribution.

In this case, we give an approach labeled mathematical deduction approach to construct bank risk data set (Fig. 2.4). This approach works because there are linkages between the known risk statistics and unknown parameters of risk distributions. Hence, we can transform the known risk statistics into parameters of risk distributions via mathematical deduction, and then determine the risk distributions. Obviously, the core issue influencing this approach is how to derive the parameters of risk distributions based on the known risk statistics.

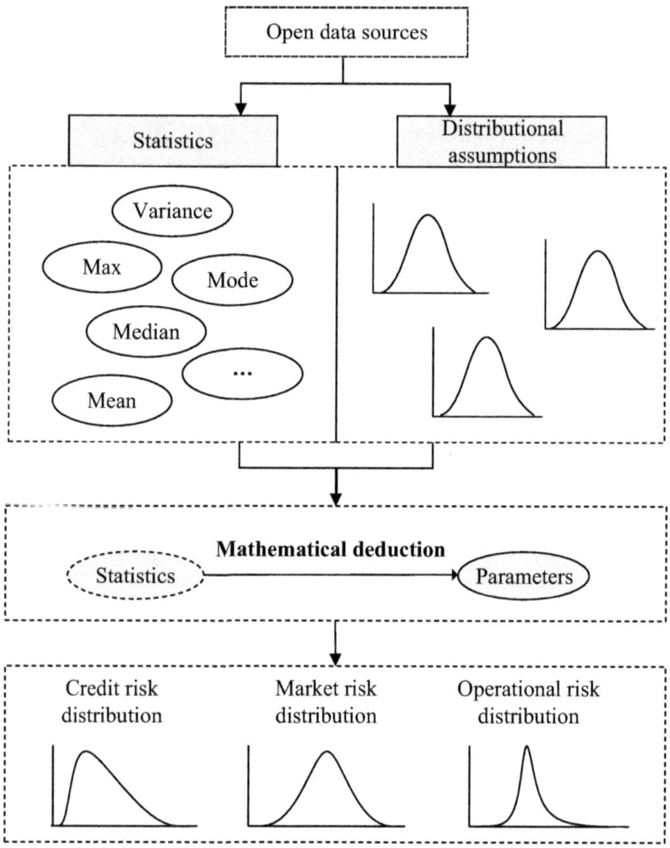

Fig. 2.4 Illustration of the mathematical deduction approach

2.4.4 Real Risk Loss Data from Open Channels

The actual risk loss data refers to the loss data of commercial banks due to the influence of various risks such as market risk, credit risk and operational risk. Public information channels, such as news reports, newspapers, literature, financial statements of listed companies, investigation reports, etc., will contain the real data of these actual risk loss events. Specifically, when bank risk events occur, radio, television, newspapers and the Internet will be widely reported. For example, when browsing newspapers or websites, you often see reports of risk events. These reports usually disclose the details of the whole risk event, such as the amount of loss, the banks involved, key personnel involved, etc.

In addition, some bank risk events have been written into the published books as bank risk management cases. Through the study of risk cases in books, we can clearly capture the context of the whole risk event. Therefore, a complete and detailed description of risk events can be obtained by collecting publicly available risk event information (Li et al. 2014a, 2014b). However, the risk loss event is a small probability event, and there is not much historical data that can be collected. Therefore, it takes a long time to build a dataset with enough data and strong reliability (Zhu et al. 2018).

At present, many commercial organizations have built external operational risk databases by collecting operational risk loss events from open channels, such as Algo OpData, Algo FIRST and SAS OpRisk Global Data (Biell and Muller 2013; Wang and Hsu 2013). SAS OpRisk Global Data is the largest external operational risk database in the world, recording more than 32,000 operational risk loss events with loss amounts exceeding the US$ 100,000 since 1831, of which 66% of operational risk events occurred in the United States (De Jongh et al. 2015). Li et al. (2014a, 2014b) constructed the largest operational risk database in China, recording 2132 operational risk events from 1994 to 2012 (Zhu et al. 2018). However, generally, only extreme loss data will be disclosed by the media, because it has reached the level of "huge loss that can not be hidden", so there will be a certain deviation. Therefore, some scholars have great controversy on using these data for research (De Fontnouvelle et al. 2006; Chaudhury 2010).

By reviewing 301 articles had been published on the topic of operational risk in banks from 2002 to March 31, 2017, we summarize the construction conditions of operational loss databases all over the world. As shown in Fig. 2.5, public databases can take three forms according to different types of providers. One form is commercial database provided by commercial vendors, including Algo OpData, Algo FIRST, SAS OpRisk GlobalData, OpBase and Willis Towers Watson (WTW) database. Another form is consortium public database provided by consortia of financial institutions, including ORX News and ÖffSchOR. The other form is self-collected public database provided by researchers, including database constructed by Solakoğlu and Köse (2009) and Chinese Operational Loss Database (COLD). In what follows, we describe these three forms of public databases in detail.

Among public databases, the majority are commercial databases provided by commercial vendors, which contain operational loss events in numerous financial

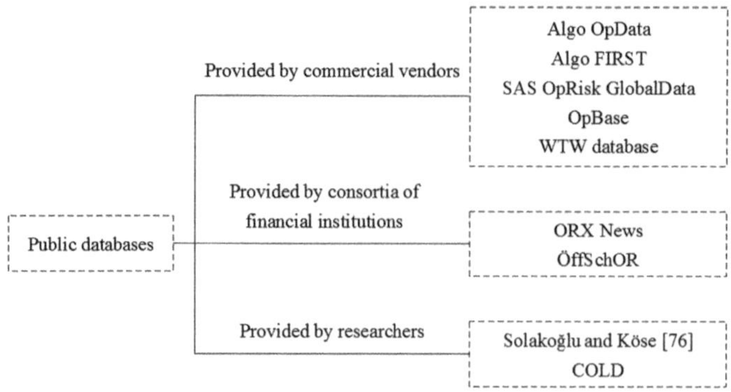

Fig. 2.5 Three forms of 9 public databases summarized from 301 articles

institutions across the world over the past decades (Wei 2006). The prominent commercial databases include Algo OpData, Algo FIRST, SAS OpRisk Global Data, WTW database and OpBase.

Besides constructing pooled industry database by requiring member financial institutions to submit internal loss data, some consortia of financial institutions, including ORIC, ORX and ÖffSchOR also develop public databases by collecting information of operational loss events available in the public domain.

For developed countries, many external operational loss databases have been constructed for potential use. However, hardly any external databases provided by consortia of banks or commercial vendors focus on developing countries' banks (Li et al. 2009). Thus, researchers in developing countries have to construct external databases by themselves to address the problem of data deficiency.

Specifically, Solakoğlu and Köse (2009) possessed a self-collected database with 22 operational loss events in the Turkish banking sector between 1998 and 2007. The COLD that constructed by Chinese researchers is the most comprehensive operational risk database in China. It contains 2132 operational loss events that occurred in the entire Chinese banking system over the years 1994 to 2012 and has already been successfully applied to several published studies (Li et al. 2014a, 2014b; Zhu et al. 2018). However, it is not easy for researchers to construct operational databases by extracting detailed information of each operational loss events from huge amounts of public available data (Gu et al. 2010).

However, there are three shortcomings in using such data for risk aggregation. First, only extreme loss data will be disclosed by the media, so the collected data have certain deviation, and some scholars have great controversy about using such data for research; second, it takes a lot of energy to collect risk loss events manually; third, it is due to risk loss events It is a small probability event, and there is not much historical data that can be collected. It takes a long time to build a data set with enough data and strong reliability (Zhu et al. 2018).

2.4.5 Financial Statement Data

The bank's risk comes from the daily business, and the bank's financial statements record the bank's operating conditions when facing various risks. Therefore, financial statements are an important data source for external personnel to study bank risks (Rosenberg and Schuermann 2006). Moreover, financial statements have attracted more and more scholars' attention due to their easy access, standardized disclosure and information aggregation, and become a major research direction to solve the problem of data missing in bank risk aggregation. A series of bank risk aggregation methods based on financial statements are proposed (Li et al. 2018).

Rosenberg and Schuermann (2006) and Li et al. (2012) pointed out that the information contained in bank financial statements is an important data source for external researchers to study bank risk. Specifically, the profit and loss account in the income statement reflects the profit and loss brought by the business activities of the bank in the face of various risks in a certain period of time. As long as we find the internal relationship between the income statement accounts and the types of risks they face, we can get the first-hand risk profit and loss data (Kuritzkes and Schuermann 2007). The bank's profit and loss is brought by the various businesses operated by the bank, so the bank's assets constitute the bank's risk exposure. Therefore, by matching the asset data in the balance sheet with the bank risk type, we can obtain the risk exposure of various risks (Li et al. 2018).

In addition to the numerical data in the income statement and balance sheet, recent studies have begun to analyze the text risk information disclosed in financial statements (Athanasakou and Hussainey 2014; Abed et al. 2016). The financial statements of Listed Companies in the United States contain text information specifically for risks. Since 2005, the securities and Exchange Commission (SEC) has required listed companies to add a new chapter "risk factor" in their financial statements, which specifically uses text to disclose the risk factors that bring speculative opportunities or risks to banks (SEC 2005).

In addition, different from the historical operating status of the bank recorded in the balance sheet and income statement, this part of text risk information contains forward-looking information and discloses the factor fluctuations that may affect the bank's risk status in the future. In recent years, text risk disclosure has become the most valuable content in financial statements (Campbell et al. 2014). Some studies began to identify risk factors by analyzing the text risk information (Huang and Li 2011; Bao and Datta 2014; Dyer et al. 2017; Miller 2017; Wei et al. 2019a, 2019b). As the disclosure of financial statements becomes more and more standardized and the contents are more and more comprehensive, collecting risk data from financial statements is a feasible method to solve the problem of lack of risk data.

Figure 2.6 illustrates how to collect risk data from open data sources. There are many open data sources, such as papers, books, newspapers, surveys and websites. The data, information or knowledge of bank risks drawn from these sources can be categorized into three major types, i.e. risk event descriptions, financial statements, risk statistics and risk distribution assumptions. If the descriptions of the whole

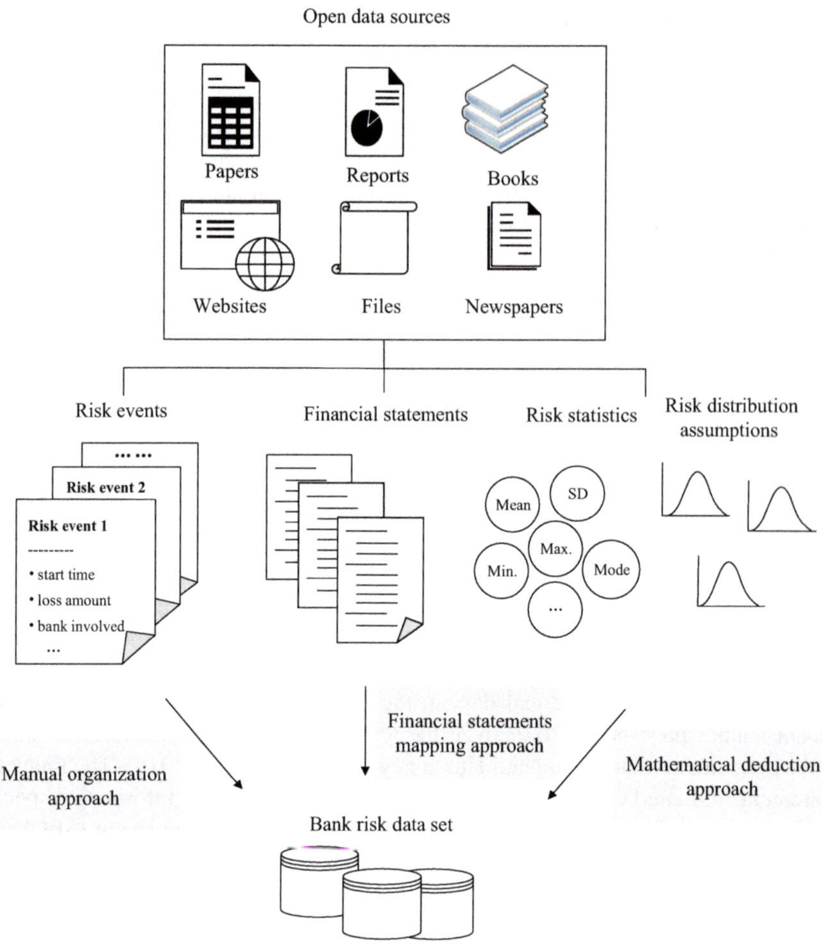

Fig. 2.6 The whole process of the general bank risk data set construction framework

risk events are available, the bank risk data set can be constructed via the manual organization approach. Besides, if there are financial statements of banks, we can construct a bank risk data set by using the financial statements mapping approach. At last, if only some risk statistics and distributional assumptions are known, we can also construct bank risk data set through the mathematical deduction approach. Bank risk data are collected based on different data types.

Figure 2.7 compares the three approaches. In particular, the manual organization approach is the most ideal because the detailed and complete information on the entire risk event can be obtained by manual collection, which enables various levels of analyses and calculations. However, the low-frequency nature of risk events implies that it will take a long time to construct a sufficiently large data set. Given that listed banks regularly disclose financial statements, the collection of risk data via

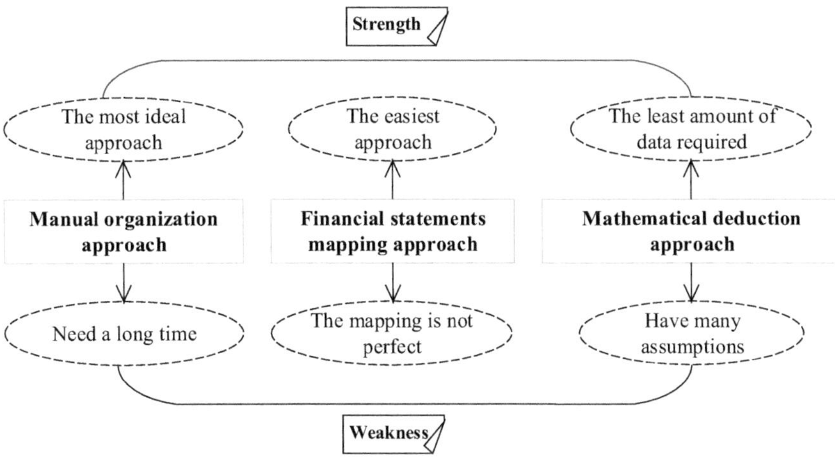

Fig. 2.7 The comparison among three approaches for bank risk dataset construction

the financial statements mapping approach is the easiest. Nevertheless, the imperfect mapping relationship hampers the reliability of risk proxies. Compared with the other two approaches, the least amount of data (risk statistics and distributional assumptions) is required by the mathematical deduction approach to establish marginal risk distributions. In turn, however, the over-reliance on risk distributional assumptions is the weakness of this approach.

2.5 Main Challenges in Bank Risk Aggregation

A major obstacle to bank risk aggregation is the problem of data availability (BCBS 2010a; Galloppo and Previati 2014). Severe events occur very infrequently, so very little historical data are available. Therefore, it takes some time before the size and quality of most institutions' databases are good enough to allow for reliable estimation of parameters (Aas et al. 2007; Cormach 2014). Many quantitative approaches typically fail due to the poor quality and low quantity of data available in the banks (Bignozzi and Tsanakas 2014; Aas et al. 2007). An obvious example is that notable data deficiencies resulted in the poor performance of risk management methods in the subprime crisis of 2007 (BCBS 2010a).

Due to the characteristics of easy access, standardized disclosure and comprehensive information, in recent years, the use of financial statement data for bank risk aggregation has become a direction to solve the problem of lack of risk data (Zhu et al. 2018). In practice, the bank's risk comes from the daily business, and the bank's financial statements record the bank's operating conditions. Therefore, the information contained in the financial statements is an important data source for bank external personnel to study bank risk (Zhu et al. 2018). In practice, the risks of the bank are rooted in day-to-day business. Financial statements present a relatively complete picture of a bank's performance, in which every earning and loss has to be recorded and summarized into an income statement periodically (Carmona and Trombetta 2008). Therefore, the income statement items are able to reflect the earnings and losses of different risk types in a relatively general way.

There have been a number of studies, see for instance Rosenberg and Schuermann (2006), Kuritzkes and Schuermann (2007) and Inanoglu and Jacobs (2009), which have obtained bank risk data by mapping income statement items into various risk types. Usually, obtaining financial statements is becoming increasingly easier, so looking for risk proxies from financial statements is a relatively easy way to derive bank risk data. Therefore, more and more scholars collect risk data from financial statements to measure overall risk, and propose a series of bank risk aggregation methods based on financial statements (Rosenberg and Schuermann 2006; Kuritzkes and Schuermann 2007; Li et al. 2018). There have been a number of studies, see for instance Rosenberg and Schuermann (2006), Kuritzkes and Schuermann (2007) and Inanoglu and Jacobs (2009) suggesting that mapping income statement items into risk types is an alternative way to collect risk data.

However, aggregating bank risks based on financial statements faces three main challenges. One is how to accurately establish the mapping relationship between bank risks and financial statements, including income statement, balance sheet, off-balance sheet and so on. The mappings between income statement items and various risk types are not identical in different studies, especially the mappings of credit risk and operational risk. Furthermore, since accounting standards are significantly different worldwide, the mapping relationship is of difference in different countries like the US and China.

Second, the disclosure frequency of financial statements is relatively low, which is generally disclosed on a quarterly basis. For the immature banking system in developing countries, the amount of risk data collected is very limited due to the small number of listed companies and short listing time, which cannot solve the problem of data shortage. Banking systems in developing countries were established relatively late and the regulations for public disclosure of risk data are not in place; the problem of data sparseness is even worse in developing countries like China. Last, financial statement data record the historical performance of banks, and there is a lag problem (Beneish et al. 2015). The results of bank risk aggregation based on historical data ignore the future trend of risk, and may not cover the loss when the future risk occurs, which puts the bank in trouble.

References

Aas K, Dimakos XK, Oksendal A (2007) Risk capital aggregation. Risk Manage 9(2):82–107. https://doi.org/10.1057/palgrave.rm.8250027

Aas K, Czado C, Frigessi A, Bakken H (2009) Pair-copula constructions of multiple dependence. Insur Math Econ 44(2):182–198. https://doi.org/10.1016/j.insmatheco.2007.02.001

Abadi MSE (2015) Analysis of new techniques for risk aggregation and dependence modelling. Doctoral dissertation, Instituto Superior de Economia e Gestão

Abed S, Al NB, Roberts C (2016) Measuring annual report narratives disclosure: empirical evidence from forward-looking information in the UK prior the financial crisis. Manage Audit J 31(4/5):338–361. https://doi.org/10.1109/ICSC.1996.864276

Alexander C, Pezier J (2003) On the aggregation of market and credit risks. ISMA Centre discussion papers in Finance No. 2003-13. University of Reading. https://doi.org/10.2139/ssrn.478381

Allen D, Faff R (2012) The global financial crisis: some attributes and responses. Acc Finan 52(1):1–7. https://doi.org/10.1111/j.1467-629X.2011.00416.x

Andrievskaya IK, Penikas HI (2012) Copula-application to modelling Russian banking system capital adequacy according to Basel II IRB-approach. Model Assist Stat Appl 7(4):267–280. https://doi.org/10.3233/MAS-2012-0239

Angela C, Bisignani R, Masala G, Masala G, Micocci M (2009) Advanced operational risk modelling in banks and insurance companies. Investment Manage Finan Innov 6(3):73–83

Arbenz P, Hummel C, Mainik G (2012) Copula based hierarchical risk aggregation through sample reordering. Insurance Math Econ 51(1):122–133. https://doi.org/10.1016/j.insmatheco.2012.03.009

Athanasakou V, Hussainey K (2014) The perceived credibility of forward-looking performance disclosures. Account Bus Res 44(3):227–259. https://doi.org/10.1080/00014788.2013.867403

Bao Y, Datta A (2014) Simultaneously discovering and quantifying risk types from textual risk disclosures. Manage Sci 60(6):1371–1391. https://doi.org/10.1287/mnsc.2014.1930

Basel Committee on Banking Supervision (2003) The 2002 loss data collection exercise for operational risk: summary of the data collected. Bank for International Settlements, Basel, Switzerland

Basel Committee on Banking Supervision (2006) International convergence of capital measurement and capital standards: a revised framework. Bank for International Settlements, Basel, Switzerland

Basel Committee on Banking Supervision (2009) Findings on the interaction of market and credit risk. Bank for International Settlements, Basel, Switzerland

Basel Committee on Banking Supervision (2010a) Basel III: a global regulatory framework for more resilient banks and banking systems. Bank for International Settlements, Basel, Switzerland

Basel Committee on Banking Supervision (2010b) Developments in modelling risk aggregation. Bank for International Settlements, Basel, Switzerland

Basel Committee on Banking Supervision (2011) Core principles for effective banking supervision. Bank for International Settlements, Basel

Baselga PL, Trujillo PA, Cardone RC (2015) Factors influencing bank risk in Europe: evidence from the financial crisis. North Am J Econ Finan 34:138–166. https://doi.org/10.1016/j.najef.2015.08.004

Bauwens L, Laurent S, Rombouts JVK (2006) Multivariate GARCH models: a survey. J Appl Econ 21(1):79–109. https://doi.org/10.1002/jae.842

Bedford T, Cooke RM (2002) Vines: a new graphical model for dependent random variables. Ann Stat 30(4):1031–1068. https://doi.org/10.2307/1558694

Beneish MD, Miller BP, Yohn TL (2015) Macroeconomic evidence on the impact of mandatory IFRS adoption on equity and debt markets. J Account Public Policy 34(1):1–27. https://doi.org/10.1016/j.jaccpubpol.2014.10.002

Berger AN, El Ghoul S, Guedhami O, Roman RA (2016) Internationalization and bank risk. Manage Sci 63(7):2283–2301. https://doi.org/10.1287/mnsc.2016.2422

Biell L, Muller A (2013) Sudden crash or long torture: the timing of market reactions to operational loss events. J Bank Finance 37(7):2628–2638. https://doi.org/10.1016/j.jbankfin.2013.02.022
Bignozzi V, Tsanakas A (2014) Parameter uncertainty and residual estimation risk. J Risk Insur 83(4):949–978. https://doi.org/10.1111/jori.12075
Bocker K, Kluppelberg C (2008) Modeling and measuring multivariate operational risk with Levy copulas. J Oper Risk 3(2):3–27. https://doi.org/10.21314/JOP.2008.043
Bouye E, Gaussel N, Salmon M (2002) Investigating dynamic dependence using Copula. Manuscript, Financial Econometrics Research Center. https://doi.org/10.2139/ssrn.1272353
Brechmann EC, Czado C, Aas K (2012) Truncated regular vines in high dimensions with application to financial data. Can J Stat 40(1):68–85. https://doi.org/10.1002/cjs.10141
Brechmann EC, Hendrich K, Czado C (2013) Conditional copula simulation for systemic risk stress testing. Insur Math Econom 53(3):722–732. https://doi.org/10.1016/j.insmatheco.2013.09.009
Brechmann E, Czado C, Paterlini S (2014) Flexible dependence modeling of operational risk losses and its impact on total capital requirements. J Bank Finance 40:271–285. https://doi.org/10.2139/ssrn.2252481
Breuer T, Jandacka M, Rheinberger K, Summer M (2008) Compounding effects between market and credit risk: the case of variable rate loans. In: Resti A (eds) Pillar II in the new Basel accord, the challenge of economic capital. Risk books, London
Breuer T, Jandacka M, Rheinberger K, Summer M (2008) Regulatory capital for market and credit risk interaction: is current regulation always conservative? Working Paper, Deutsche Bundesbank
Brockmann M, Kalkbrener M (2010) On the aggregation of risk. J Risk 12(3):45–68. 10.21314.JOR.2010.208
Bruneton JP (2011) Copula-based hierarchical aggregation of correlated risks. In: The behaviour of the diversification benefit in Gaussian and Lognormal trees. http://arxiv.org/abs/1111.1113
Bürgi R, Dacorogna M, Iles R (2008) Risk aggregation, dependence structure and diversification benefit. In: Rösch D, Scheule H (eds) Stress-testing for financial. https://doi.org/10.2139/ssrn.1468526
Busetti F, Harvey A (2010) When is a copula constant? A test for changing relationships. J Finan Econ 9(1):106–131. https://doi.org/10.1093/jjfinec/nbq020
Campbell JL, Chen HC, Dhaliwal DS, Lu HM, Steele LB (2014) The information content of mandatory risk factor disclosures in corporate filings. Rev Acc Stud 19(1):396–455. https://doi.org/10.1007/s11142-013-9258-3
Carmona S, Trombetta M (2008) On the global acceptance of IAS/IFRS accounting standards: the logic and implications of the principles-based system. J Account Public Policy 27(6):455–461. https://doi.org/10.1016/j.jaccpubpol.2008.09.003
Chaudhury M (2010) A review of the key issues in operational risk capital modeling. J Oper Risk 5(3):37–66. https://doi.org/10.21314/JOP.2010.078
Chen YH, Tu AH (2013) Estimating hedged portfolio value-at-risk using the conditional copula: an illustration of model risk. Int Rev Econ Finan 27:514–528. https://doi.org/10.1016/j.iref.2013.01.006
Cherubini U, Luciano E, Vecchiato W (2004) Copula methods in finance. Wiley, Chichester. https://doi.org/10.1002/9781118673331.ch8
Chollete L, Heinen A, Valdesogo A (2009) Modeling international financial returns with a multivariate regime-switching copula. J Finan Econ 7(4):437–480. https://doi.org/10.2139/ssrn.1102632
Cormach CM (2014) Fitting operational risk data using limited information below the threshold. J Oper Risk 9(2):39–56. https://doi.org/10.21314/JOP.2014.142
Corrigan J, De Decker J, Hoshino T, Van Delft L, Verheugen H (2009) Aggregation of risks and allocation of capital. Milliman
Darbellay G (1998) An adaptive histogram estimator for the mutual information. UTIA Research Report no. 1889. Prague Academy of Science

De Fontnouvelle P, Dejesus-Rueff V, Jordan J, Rosengren E (2006) Capital and risk: new evidence on implications of large operational losses. J Money Credit Bank 38(7):1819–1846. https://doi.org/10.1353/mcb.2006.0088

De Jongh R, De Wet T, Raubenheimer H, Venter JH (2015) Combining scenario and historical data in the loss distribution approach: a new procedure that incorporates measures of agreement between scenarios and historical data. J Oper Risk 10(1):45–76. https://doi.org/10.21314/JOP.2015.160

Dimakos XK, Aas K (2004) Integrated risk modelling. Stat Model 4(4):265–277. https://doi.org/10.1191/1471082X04st079oa

Dionisio A, Menezes R, Mendes DA (2006) Entropy-based independence test. Nonlinear Dyn 44(1–4):351–357. https://doi.org/10.1007/s11071-006-2019-0

Drehmann M, Sorensen S, Stringa M (2010) The integrated impact of credit and interest rate risk on banks: a dynamic framework and stress testing application. J Bank Finance 34(4):713–729. https://doi.org/10.1016/j.jbankfin.2009.06.009

Dyer T, Lang M, Stice LL (2017) The evolution of 10-K textual disclosure: evidence from latent dirichlet allocation. J Account Econ 64(2–3):221–245. https://doi.org/10.2139/ssrn.2741682

El-Gamal M, Inanoglu H, Stengel M (2007) Multivariate estimation for operational risk with judicious use of extreme value theory. J Oper Risk 2(1):21–54. https://doi.org/10.1111/j.0022-2879.2007.00043.x

Elsinger H, Lehar A, Summer M (2006) Risk assessment for banking systems. Manage Sci 52(9):1301–1314. https://doi.org/10.2139/ssrn.423985

Embrechts P, McNeil A, Straumann D (1999) Correlation: pitfalls and alternatives. Risk 12(5):69–71

Embrechts P, Furrer H, Kaufmann R (2003) Quantifying regulatory capital for operational risk. Deriv Use Trading Regul 9(3):217–233

Engle R (2002) Dynamic conditional correlation: a simple class of multivariate generalized autoregressive conditional heteroskedasticity models. J Bus Econ Stat 20(3):339–350. https://doi.org/10.1198/073500102288618487

Figini S, Giudici P, Uberti P (2010) A threshold based approach to merge data in financial risk management. J Appl Stat 37(11):1815–1824. https://doi.org/10.1080/02664760903164921

Gaisser S, Memmel C, Schmidt R, When CS (2011) Time dynamic and hierarchical dependence modeling of a supervisory portfolio of banks: a multivariate nonparametric approach. J Risk 14(1):3–40. https://doi.org/10.21314/JOR.2011.233

Galloppo G, Previati D (2014) A review of methods for combining internal and external data. Oper Risk, 9(4):83–103. https://doi.org/10.21314/JOP.2014.135

Garcia R, Tsafack G (2008) Dependence structure and extreme comovements in international equity and bond markets with portfolio diversification effects. Working paper. EDHEC Risk Asset Management Research Centre. https://doi.org/10.2139/ssrn.1107718

Greenwood R, Landier A, Thesmar A (2015) Vulnerable Banks. J Finan Econ 3:471–485

Grundke P (2010) Top-down approaches for integrated risk management: how accurate are they? Eur J Oper Res 203(3):662–672. https://doi.org/10.1016/j.ejor.2009.09.015

Grundke P (2013) On the reliability of integrated risk measurement in practice. J Risk 15(3):87–110. https://doi.org/10.21314/JOR.2013.260

Guegan D, Maugis PA (2009) Prospects on vines. Working paper

Guegan D, Hassani BK (2012) Multivariate VaRs for operational risk capital computation: a Vine structure approach. Documents De Travail Du Centre Deconomie De La Sorbonne 17(2):235–246. https://doi.org/10.1504/IJRAM.2013.057104

Guegan D, Zhang J (2010) Change analysis of a dynamic copula for measuring dependence in multivariate financial data. Quant Finan 10(4):421–430. https://doi.org/10.1080/14697680902933041

Gu J, Zhang L (2010) Some comments on big data and data science. Annals of data science, 1(3), 283–291. https://doi.org/10.1007/s40745-014-0021-9

Hafner CM, Manner H (2012) Dynamic stochastic copula models: estimation, inference and applications. J Appl Econ 27(2):269–295. https://doi.org/10.1002/jae.1197

Hafner CM, Reznikova O (2010) Efficient estimation of a semiparametric dynamic copula model. Comput Stat Data Anal 54(11):2609–2627. https://doi.org/10.1016/j.csda.2010.01.013
Hartmann P (2010) Interaction of market and credit risk. J Bank Finance 34(4):697–702. https://doi.org/10.1016/j.jbankfin.2009.10.013
Hofmann, M., & Czado, C. (2010). Assessing the VaR of a portfolio using D-vine copula based multivariate GARCH models. Working paper
Hu L (2006) Dependence patterns across financial markets: a mixed copula approach. Appl Finan Econ 16(10):717–729. https://doi.org/10.1080/09603100500426515
Huang KW, Li ZL (2011) A multilabel text classification algorithm for labeling risk factors in SEC form 10-K. ACM Trans Manage Inf Syst (TMIS) 2(3):1–19. https://doi.org/10.1145/2019618.2019624
Huang M, Wu CC (2015) Economic benefits and determinants of extreme dependences between REIT and stock returns. Rev Quant Finan Acc 44(2):299–327. https://doi.org/10.1007/s11156-013-0407-3
Huang X, Zhou H, Zhu H (2012) Assessing the systemic risk of a heterogeneous portfolio of banks during the recent financial crisis. J Finan Stab 8(3):193–205. https://doi.org/10.2139/ssrn.1459946
Hull JC (2012) Risk management and financial institutions, 3rd ed. Wiley, Hoboken. http://tainguyenso.vnu.edu.vn/jspui/handle/123456789/51683
Imbierowicz B, Rauch C (2014) The relationship between liquidity risk and credit risk in banks. J Bank Finance 40:242–256. https://doi.org/10.1016/j.jbankfin.2013.11.030
Inanoglu H, Jacobs M (2009) Models for risk aggregation and sensitivity analysis: an application to bank economic capital. J Risk Finan Manage 2(1):118–189. https://doi.org/10.2139/ssrn.1433065
International Financial Risk Institute Foundation and Chief Risk Officer Forum (2007) Insights from the joint IFRI/CRO forum survey on economic capital practice and applications. KMPG Business Advisory Services
Jarrow RA, Turnbull SM (2000) The intersection of market and credit risk. J Bank Finance 24(1–2):271–299. https://doi.org/10.1016/S0378-4266(99)00060-6
Joe H (1996) Families of m-variate distributions with given margins and m(m-1)/2 bivariate dependence parameters. Lect Notes Monograph Ser 28:120–141. https://doi.org/10.1214/lnms/1215452614
Karimalis EN, Nomikos NK (2017) Measuring systemic risk in the European banking sector: a copula CoVaR approach. Eur J Finan 1:1–38. https://doi.org/10.1080/1351847X.2017.1366350
Kaufman GG, Scott KE (2003) What is systemic risk, and do bank regulators retard or contribute to it? Independent Rev 7(3):371–391
Kiss HJ, Homolya D (2014) On the optimal design of operational risk data consortia. J Oper Risk 9(1), 33–57. https://doi.org/10.2139/ssrn.2289906
Kuritzkes A, Schuermann T (2007) What we know, don't know and can't know about bank risk: a view from the trenches. In: Diebod FX et al. (eds) The known, the unknown and the unknowable in financial risk management. Princeton University Press, Princeton
Li J, Feng J, Chen J (2009) A piecewise-defined severity distribution-based loss distribution approach to estimate operational risk: evidence from Chinese national commercial banks. Int J inf technol decis making, 8(4), 727–747. https://doi.org/10.1142/S0219622009003727
Li J, Feng J, Sun X, Li M (2012) Risk aggregation mechanisms and approaches in banking industry. Int J Inf Technol Decis Mak 11(6):1183–1213. https://doi.org/10.1142/S0219622012500320
Li J, Zhu X, Xie Y, Chen J, Gao L, Feng J, Shi W (2014a) The mutual-information based variance-covariance approach: an application to operational risk aggregation in Chinese banking. J Oper Risk 9(3):3–19. https://doi.org/10.21314/JOP.2014.140
Li L, Yuen KC, Yang J (2014b) Distorted mix method for constructing copulas with tail dependence. Insur Math Econ 57(3):77–89. https://doi.org/10.1016/j.insmatheco.2014.05.002
Li J, Zhu X, Wu D, Lee CF, Feng J, Shi Y (2015) On the aggregation of credit, market and operational risks. Rev Quant Finan Acc 44(1):161–189. https://doi.org/10.2139/ssrn.478381

Li J, Wei L, Lee CF, Zhu X, Wu D (2018) Financial statements based bank risk aggregation. Rev Quant Finan Acc 50(3):673–694. https://doi.org/10.1007/s11156-017-0642-0
Malevergne Y, Sornette D (2006) Extreme financial risks: from dependence to risk management. Springer, Berlin
McNeil A, Frey R, Embrechts P (2005) Quantitative risk management: concepts, techniques, tools. Princeton University Press, Princeton. https://doi.org/10.1198/jasa.2006.s156
Miller GS (2017) Discussion of the evolution of 10-K textual disclosure: evidence from latent dirichlet allocation. J Account Econ 64(2–3):246–252. https://doi.org/10.1016/j.jacceco.2017.07.004
Mittnik S, Paterlini S, Yener T (2013) Operational risk dependencies and the determination of risk capital. J Oper Risk 8(4):83–104. https://doi.org/10.21314/JOP.2013.133
Moddemeijer R (1999) A statistic to estimate the variance of the histogram-based mutual information estimator based on dependent pairs of observations. Signal Process 75(1):51–63. https://doi.org/10.1016/S0165-1684(98)00224-2
Morone M, Cornaglia A, Mignola G (2007) Economic capital assessement via copulas aggregation and allocation of different risk types. Working paper. IntesaSanpaolo
Nelsen RB (1999) An introduction to copulas. Springer, New York. https://doi.org/10.2307/1271100
Patton AJ (2001) Estimation of copula models for time series of possibly different. Working paper. University of California, San Diego
Patton AJ (2006a) Estimation of multivariate models for time series of possibly different lengths. J Appl Econ 21(2):147–173. https://doi.org/10.1002/jae.865
Patton AJ (2006b) Modeling asymmetric exchange rate dependence. Int Econ Rev 47(2):527–556. https://doi.org/10.1111/j.1468-2354.2006.00387.x
Pelletier D (2006) Regime switching for dynamic correlations. J Econometrics 131:445–473. https://doi.org/10.1016/j.jeconom.2005.01.013
Pérignon C, Smith DR (2010) Diversification and value-at-risk. J Bank Finance 34(1):55–66. https://doi.org/10.1016/j.jbankfin.2009.07.003
Pourkhanali A, Kim JM, Tafakori L, Fard FA (2016) Measuring systemic risk using vine-copula. Econ Model 53:63–74. https://doi.org/10.1016/j.econmod.2015.11.010
Rachev ST, Menn C, Fabozzi FJ (2005) Fat-tailed and skewed asset return distributions: implications for risk management, portfolio selection, and option pricing. Wiley, Hoboken
Rodriguez JC (2007) Measuring financial contagion: a Copula approach. J Empirical Finan 14(3):401–423. https://doi.org/10.1016/j.jempfin.2006.07.002
Rosenberg JV, Schuermann T (2006) A general approach to integrated risk management with skewed, fat-tailed risks. J Finan Econ 79(3):569–614. https://doi.org/10.1016/j.jfineco.2005.03.001
Securities and Exchange Commission (SEC) (2005) Securities and exchange commission final rule, release no. 33-8591(FR-75). Retrieved from http://www.sec.gov/rules/final/33-8591.pdf
Sklar A (1959) Fonctionde repartition a dimension stleurs marges. Publ Inst Stat Univ Paris 8:229–231
Solako˘glu MN, Köse A (2009) Operational risk and stock market returns: evidence from Turkey. In: Gre-goriou G (ed) Operational risk toward Basel III: best practices and issues in modeling, management,and regulation. Wiley finance series. Wiley, Hoboken, pp 115–128
Tang A, Valdez EA (2009) Economic capital and the aggregation of risks using copulas. Working paper. University of New South Wales. https://doi.org/10.2139/ssrn.1347675
Upper C, Worms A (2004) Estimating bilateral exposures in the German interbank market: is there a danger of contagion? Eur Econ Rev 48(4):827–849. https://doi.org/10.1016/j.euroecorev.2003.12.009
Wang TW, Hsu C (2013) Board composition and operational risk events of financial institutions. J Bank Finance 37(6):2042–2051. https://doi.org/10.1016/j.jbankfin.2013.01.027
Ward J, Lee C (2002) A review of problem-based learning. J Family Consumer Sci Educ 20 (1), 16-26. http://www.natefacs.org/JFCSE/v20no1/v20no1.htm

Wei L, Li J, Zhu X (2018) Operational loss data collection: a literature review. Annals of Data Science, 5, 313–337. https://doi.org/10.1007/s40745-018-0139-2
Wei L, Li G, Zhu X, Li J (2019a) Discovering bank risk factors from financial statements based on a new semi-supervised text mining algorithm. Acc Finan, In press. https://doi.org/10.1111/acfi.12453
Wei L, Li G, Zhu X, Sun X, Li J (2019b) Developing a hierarchical system for energy corporate risk factors based on textual risk disclosures. Energy Econ 80:452–460. https://doi.org/10.1016/j.eneco.2019.01.020
Weiß GNF (2013) Copula-GARCH versus dynamic conditional correlation: an empirical study on VaR and ES forecasting accuracy. Rev Quant Finan Acc 41(2):179–202. https://doi.org/10.1007/s11156-012-0311-2
Wei R (2006) Quantification of operational losses using firm-specific information and external databases. Journal of Operational Risk, 1(4), 3–34. https://doi.org/10.21314/JOP.2007.017
Yang S, Li J, Cai J, Guo K, Gao X, Meng F (2014) Dataoriented method to big data standard system creation: a case of Chinese financial industry. Annals Data Sci (3–4), 325–338. https://doi.org/10.1007/s40745-014-0024-6
Yao Y, Li J, Zhu X, Wei L (2017) Expected default based score for identifying systemically important banks. Econ Model 64:589–600. https://doi.org/10.1016/j.econmod.2017.04.023
Yun J, Moon H (2014) Measuring systemic risk in the Korean banking sector via dynamic conditional correlation models. Pac Basin Finan J 27:94–114. https://doi.org/10.1016/j.pacfin.2014.02.005
Zhu X, Wei L, Wu D, Li J (2018) A general framework for constructing bank risk data sets. J Risk 21(1):37–59. https://doi.org/10.21314/JOR.2018.393

Chapter 3
Financial Statements-Based Bank Risk Aggregation Framework

3.1 The Proposed Framework of Financial Statements-Based Bank Risk Aggregation

Figure 3.1 illustrates the existing financial statements-based bank risk aggregation framework. Specifically, Chap. 4 describes the bank risk aggregation approach based on the income statement. By mapping different bank risk types into income statement items, risk profit & loss data are obtained to measure bank risks. In the empirical analysis, the bank risk aggregation approach based on the income statement is adopted to aggregate credit, market, liquidity, and operational risks of the Chinese banking system.

The disclosure of financial statements is low frequency (usually issued quarterly), which may lead to data shortage, especially in developing countries. Thus, Chap. 5 focuses on the low-frequency problem of financial statements data. In this chapter, a novel high-frequency-factor-integral bank risk aggregation approach is proposed. By transforming the aggregation of low-frequency risk data into the integration of high-frequency risk factors, the problem of low-frequency data in financial statements is solved. The empirical research in this chapter uses a factor-integral approach to aggregate credit risk and market risk of Chinese listed banks. Further, it compares the results with the aggregation results based on low-frequency risk data to verify the effectiveness of the proposed approach.

The factor-integral approach proposed in Chap. 5 can solve the low-frequency of financial statements data based on high-frequency risk factor data. However, it is not appliable to risks without common risk factors. Thus, to solve the data shortage of risks without common risk factors, some other data sources are needed to supplement financial statements data to arrive at the risk aggregation of multiple bank risks with different characteristics. Chapter 6 propose a two-stage risk aggregation approach to aggregate credit, market and operational risks based on financial statements and external loss database. Specifically, in the first stage, mapping credit and market risks that have common risk factors into income statements to obtain risk profit and loss data. Then by establishing a mapping relationship between risk profit & loss and

J. Li et al., *Financial Statements-Based Bank Risk Aggregation*, Innovation in Risk Analysis, https://doi.org/10.1007/978-981-19-0408-0_3

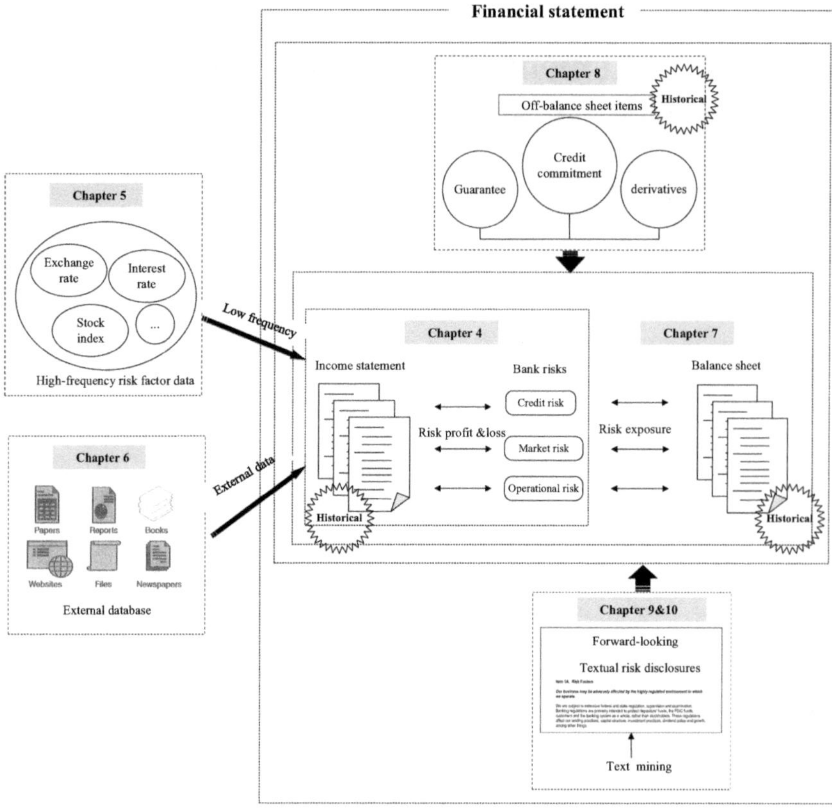

Fig. 3.1 The existing financial statements-based bank risk aggregation framework

the underlying high-frequency risk factors through the factor-integral approach, we obtain the aggregate risk of credit and market risk in the first stage based on financial statements data. In the second stage, operational risk without common risk factors is measured based on an external loss database. And then using the hierarchical copula to aggregate the operational risk and the aggregate risk obtained in the first stage to arrive at the total risk. The proposed two-stage risk aggregation approach is empirically compared with three commonly used risk aggregation approaches, including simple summation approach, variance/covariance approach and multivariate copula approach by applying them to the Chinese banking system to aggregate credit, market and operational risks.

The above three Chapters collect risk profit & loss data from income statements. Chapter 7 describes the bank risk aggregation approach based on the income statement and balance sheet by fully using income statement and balance sheet data. Specifically, by mapping income statements items into different bank risk types, we can obtain risk profit & loss data. By mapping balance sheet items and different risk types, risk exposure is obtained. In the empirical analysis, credit, market, liquidity and operational risks of the Chinese banking system are aggregated using the bank risk aggregation approach based on the income statement and balance sheet.

Besides on-balance sheet assets, off-balance sheet activities bring additional risks. Off-balance sheet items are also risking exposure that can result in risk profit & loss. Ignoring off-balance sheet activities in bank risk aggregation may lead to biased total risk results. In Chap. 8, the off-balance sheet items are incorporated into the bank risk aggregation. By establishing the mapping relationship between the risk types and the income statement, the assets on and off the balance sheet, the total bank risks brought by the on balance sheet business and the off-balance sheet business are measured completely. The empirical research in this chapter integrates the credit risk, market risk, liquidity risk and operational risk of Chinese listed banks, calculates the total risk brought by off-balance sheet business in the Chinese banking industry. Moreover, it empirically analyzes the impact of off-balance sheet items on bank risk and whether the impact is related to bank size.

The main drawback of numerical data recorded in financial statements is the existence of hysteresis (Beneish et al. 2015). Bank risk aggregation based on historical numerical financial statement data is less timely. Chapters 9 and 10 incorporate forward-looking information disclosed in financial statements into bank risk aggregation to overcome the hysteresis of historical numerical financial statement data. Specifically, Chap. 9 comprehensively and accurately identifies bank risk factors by proposing a new semi-supervised text mining naive collision algorithm to analyze the forward-looking textual risk factor disclosure reported in financial statements of the US banking system. Then in Chap. 10, the forward-looking textual risk disclosures are incorporated into bank risk aggregation by mapping the identified bank risk factors into different risk types. Using the historical data and forward-looking risk information in financial statements, the total risk results from risk aggregation can reflect the historical situation and future trend of risk, which can enhance the timeliness of risk aggregation results. The empirical research in this chapter integrates the credit risk, market risk and operational risk of American listed banks. Further, it analyzes the impact of forward-looking risk information on the results of bank risk aggregation.

Chapter 11 describes the main research work of this book, summarizes the main research results, and introduces the future research direction.

3.2 The Basic Idea of Financial Statements Based Bank Risk Aggregation

The basic idea of the risk aggregation framework based on financial statements proposed in this book can be described as follows:

(1) Mapping marginal risk and integrated risk into their financial statement components;
(2) Collecting and calculating returns of earnings for related risks;

(3) Measuring the marginal risks
(4) Quantifying textual forward-looking risk information
(5) Aggregating marginal risk using risk aggregation approaches with forward-looking risk information.

Specifically, the risk is usually understood in two ways: one is to emphasize the uncertainty of return; the other is to emphasize the uncertainty of loss. According to the narrow definition of risk, risk can only show loss, and there is no possibility of profit from risk. The broad definition of risk holds that the result of risk may bring loss, profit, or no loss and no profit. Financial risk belongs to broad sense risk.

In the traditional risk measurement, loss is often taken as the research object, which belongs to the narrow sense of risk. On the one hand, if the loss data cannot be obtained, the risk measurement method using loss data may not be available; on the other hand, this narrow sense risk view which only considers the loss uncertainty is not comprehensive. Mathematically, risk is usually defined as the expected change in the outcome of one or more future events. In particular, financial risk is usually defined as the expected change in the price of an asset or its profit or loss. In banking, risk is usually defined as the fluctuation of profit and loss. Therefore, the measurement of financial risk should consider the broad sense of risk.

The financial statement is an important way to obtain the operation condition of enterprises taking a variety of risks. Financial statements, also known as external accounting statements, are the accounting statements provided by the accounting entity to reflect the financial status and operation of the accounting entity. They generally include balance sheets, income statements, cash flow statements, notes to financial statements. The financial statement is a comprehensive reflection of the business performance and financial status of an enterprise in a certain period; at the same time, it also reflects the risks that the enterprise undertakes to a certain extent. In financial management, through the analysis of financial statements disclosed by enterprises, we can obtain the risk status of enterprises to a certain extent. Therefore, the financial statement is a meaningful way to reveal enterprise risk information.

Like other enterprises, the financial statements of financial enterprises contain the information of their operating performance and financial status in a certain period, especially the various risks they bear, and the impact of these risks on assets, liabilities, profits and cash flow. Among financial statements, the income statement (or profit and loss statement) directly records the total profit and loss of financial enterprises in a certain period. Moreover, it contains various assets and liabilities affected by different types of risks and the profits and losses of various businesses. Therefore, by analyzing the items recorded in the income statements, we can obtain the profit and loss data of risks faced by financial enterprises.

Assets recorded in off-and-on balance sheet assets are exposures of banks to produce profits and losses. Especially off-balance sheet (OBS) activities, which often be ignored in research, had triggered additional risks while bringing considerable income. The role of OBS items in systemic vulnerability was highlighted during the subprime crisis. As early as 1988, the business scope under supervision had already extended from balance sheet items to OBS items (BCBS 1988). The China banking regulatory commission (CBRC) also published a policy document titled *Risk Management Guidelines of Commercial Banks' off-balance Sheet Business* to regulate OBS activities in 2011. Thus, by mapping different risk types into off-and on-balance sheet assets, risk exposures of each risk type can be obtained.

However, the main drawback of numerical data recorded in financial statements, either in income statements or balance sheets, is the existence of hysteresis (Beneish et al. 2015). Backward-looking numerical financial statement data used for bank risk aggregation only describe historical bank risk profiles and cannot reflect future conditions of bank risk. Changes in macroeconomic or political conditions may make bank risks differing as to their historical volatility characteristics (Kupiec and Ramirez 2013; Jiménez et al. 2013). Thus, bank risk aggregation based on historical numerical financial statement data is less timely. In other words, using historical data while disregarding changes in future market movements in risk aggregation can generate biased aggregation results and lead to inadequate capital against potential total losses. As discussed above, one common weakness of all the financial statements-based bank risk aggregation approaches based on historical data is that the future trends of bank risks are ignored.

Hence, forward-looking information is needed to be incorporated into bank risk aggregation to overcome the hysteresis of historical numerical financial statement data. Besides backward-looking information, forward-looking information is also disclosed in financial statements (Kılıç and Kuzey 2018). In particular, the SEC has emphasized that investors had a greater need for forward-looking disclosures than for disclosures about past events, and it has issued guidelines for companies to present any known trends, plans, and uncertainties that are likely to materially affect future operations (Muslu et al. 2014). Since 2005, textual risk factors disclosed in Section 1A of Securities and Exchange Commission (SEC) Form 10-K have offered forward-looking risk information to discuss "the most significant factors that make the offering speculative or risky" (SEC 2005). Unlike numerical data recorded in financial statements, which primarily summarize the historical operating performance of a bank, textual risk disclosures include information on risks arising from future market movements. Thus, textual risk disclosures are forward-looking sources of bank risk information that provide a detailed description of bank risk losses that may occur in the future (Zhu et al. 2016).

Based on the above analysis, this book proposes the following risk aggregation ideas: by analyzing the relationship between different risk types and risk information, including numerical historical and textual forward information recorded in financial statements, establishing the corresponding relationship between financial statement data and various risks, and using corresponding risk measurement methods to measure various risks, such as market risk, credit risk and operational risk. In this book, the above-mentioned risk aggregation idea is called "bank risk aggregation framework based on financial statements data".

3.3 The General Procedure of Financial Statements Based Bank Risk Aggregation

As shown in Fig. 3.2, the whole process of aggregating bank risks using financial statements data is given. Firstly, we use historical data recorded in off-and-on balance sheet and income statements to measure bank risks. Then we adopt techniques for text analysis to analyze textual risk disclosures recorded in financial statements to construct forward-looking adjustment index (FLAI). Finally, we use aggregation approaches to aggregate bank risks based on marginal risk results and FLAI.

3.3.1 The Steps to Use Historical Financial Statements Data

By mapping P&L items recorded in income statements into different risk types, we can get risk P&L data. By mapping off-and-on balance sheet assets into different risk types, we can obtain risk exposure data.

However, Risk P&L items from income statements are not comparable among different banks because banks differ in scale, capital allocation, investment strategy and management level (Rosenberg and Schuermann 2006). To allow direct comparison across banks, risk P&L needs to be converted into a "risk return" based measure. Following Kretzschmar et al. (2010), we use the data preprocessing method to obtain a specific bank's risk return. The procedure of data preprocessing can be divided into the following three steps:

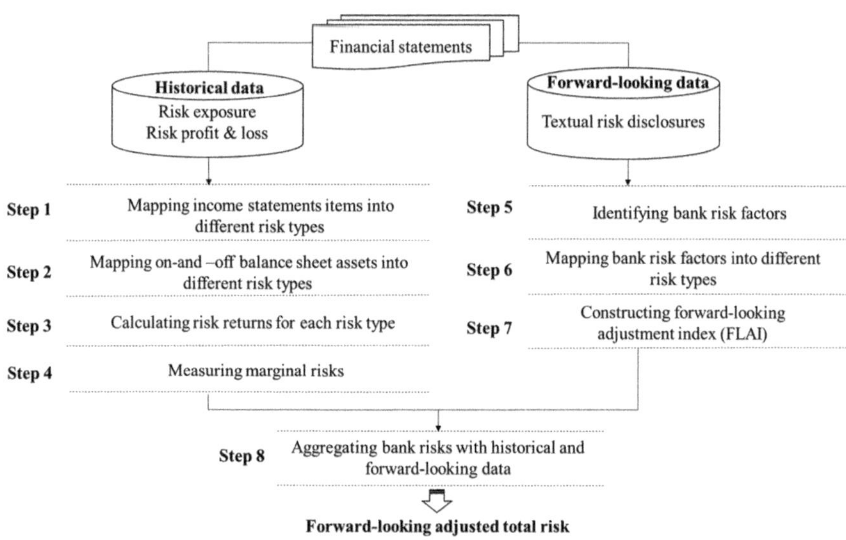

Fig. 3.2 The whole process of aggregating bank risks using financial statements data

Firstly, we convert risk P&L into a "risk return" based measure. Since risk P&L is generated by assets exposed to risk, an obvious approach would be to divide risk P&L by assets to yield a return on assets measure. In this book, bank assets are defined as risk exposures. Thus, the risk return is the ratio of risk P&L to risk exposure. We then define the marginal risk return for the ith bank, jth risk in period t as

$$r_{i,j,t} = \left(\frac{R_{i,j,t}}{RE_{i,j,t}}\right) \tag{3.1}$$

where $r_{i,j,t}$, $R_{i,j,t}$ and $RE_{i,j,t}$ stand for the risk return, risk P&L and risk exposure of bank i, risk j in period t, respectively.

Then, we compute the expected risk return and deviation from risk return. The risk return can be divided into two parts: the expected risk return and deviation from risk return. The expected risk return for a bank is the average risk return over the sample period, which reflects the bank's characteristics in terms of scale, capital allocation, investment strategy and management level. The deviation from risk return is computed by subtracting the average risk return over the sample period (expected risk return) for each bank, reflecting the macroeconomic background and operating conditions of the whole banking industry. Thus, a bank's risk return is determined by both market and individual information. Specifically, the expected risk return is defined as

$$\overline{r}_{i,j} = \left(\frac{1}{T_i}\sum_{t=1}^{T_i} r_{i,j,t}\right) \tag{3.2}$$

and deviation from risk return as

$$\Delta_{i,j,t} = r_{i,j,t} - \overline{r}_{i,j} \tag{3.3}$$

where bank i is observed for T_i periods. $\overline{r}_{i,j}$ denotes the expected risk return for bank i and risk j over the sample period T_i. $\Delta_{i,j,t}$ denotes the deviation from risk return of bank i, risk j in period t.

Finally, we obtain a typical bank's risk returns to model marginal risk distributions. For a typical bank, its risk return is determined by market information and individual information. The market information is composed of all sample banks' deviation from risk return. Thus, by combining all sample banks' deviation from risk return and the typical bank's ($i = k$) expected risk return, we finally compute a typical bank's risk return. Specifically, the typical bank's risk return is written as

$$r_{k,j,t} = \overline{r}_{k,j} + \Delta_{j,t} = \sum_i \Delta_{i,j,t} \tag{3.4}$$

where $r_{k,j,t}$ is the risk return of bank k, risk j in period t. $\overline{r}_{j,t}$ stands for the expected risk return of bank k and risk j. $\Delta_{j,t}$ denotes the summation of deviation from risk return of risk j in period t of all sample banks, reflecting the market information of risk j in period t.

Value-at-Risk (VaR), which has become a standard model for measuring and assessing risk, is used to measure marginal risk in this book (Huang 2013; Hsu et al. 2012). VaR is defined as a quantile of the distribution of risk returns. Therefore, the larger negative value or smaller positive value of VaR corresponds to the higher level of risk (Rosenberg and Schuermann 2006). However, VaR does not satisfy the subadditivity condition, so it is not a coherent risk measure (Artzner et al. 1999). A related statistic, expected shortfall (ES), which is also referred to as Conditional VaR, is a coherent risk measure that estimates the mean of the beyond VaR tail region (Rosenberg and Schuermann 2006). Thus, besides using VaR to discuss the empirical results, we conduct robustness checks using ES as well.

VaR is defined as a quantile of the distribution of risk returns (Liu and Ralescu 2017). So the larger negative value or smaller positive value of VaR corresponds to the higher level of risk (Rosenberg and Schuermann 2006). VaR at a specific confidence level $1 - \alpha$ is defined as the smallest number l such that the probability of loss L exceeding l is not larger than α:

$$VaR(\alpha) = \inf\{l : P(L \geq l) \leq \alpha\} \tag{3.5}$$

Using Eq. (3.6), ES at a specific confidence level $1 - \alpha$ is defined as the mean of the loss L exceeding VaR(α):

$$ES(\alpha) = E[L|L > VaR(\alpha)] \tag{3.6}$$

Similar to VaR, we take ES as a tail expectation of losses, thus ES is also negative.

3.3.2 *The Steps to Use Forward-Looking Textual Risk Disclosures*

Textual risk disclosures reported in Form 10-K, which consists of a summary heading and detailed explanations, include forward-looking information used to foresee risk factors exposed by banks. We map identified bank risk factors into different risk types. In previous studies, through analysis of Form 10-K textual risk disclosures, Huang and Li (2011), Bao and Datta (2014) and Campbell et al. (2014) also identified various risk factors and labeled them based on their judgments. In this book, we further establish a mapping relationship between identified risk factors and risk types to link the textual risk disclosures into different bank risk types. As individual risk is the basis for further bank risk aggregation, the mapping relationship between risk factors and risk types essentially establishes a link between textual risk disclosures and bank risk aggregation.

Bank risks are affected by certain risk factors (Jarrow and Turnbull 2000; Baselga-Pascual et al. 2015). For example, the market risk arises from adverse movements in market factors such as interest rates, exchange rates and equity prices (Hartmann 2010). As the changes in risk factors will result in bank losses (Breuer et al. 2010; Grundke 2010), some researchers have used risk factors to explain risk loss data (Rosenberg and Schuermann 2006). Thus, it is feasible and reasonable to establish a mapping relationship between identified bank risk factors and different risk types.

The mapping relationship between bank risk factors and risk types is based on definitions of bank risk factors and bank risks. The definition of each risk factor is obtained from the first stage. Bank risks have been defined by the BCBS and in several previous works (BCBS 2006; Li et al. 2015). Thus, a reasonable relationship between bank risk factors and bank risks can be established by analyzing definitions of bank risk factors and bank risks. In establishing a mapping relationship between bank risk factors and forms of risk, we can determine the annual disclosure frequency of each type of risk.

Then we construct the FLAI based on the annual disclosure frequency of each risk type to quantify textual risk disclosures. Indeed, the more frequently a given type of risk is disclosed, the more attention paid by commercial banks to this risk type, further showing that a given risk type will become more severe in the forthcoming future. Thus, the FLAI was constructed based on the principle that the annual disclosure frequency of each risk type can reflect foresight on risk severity.

Specifically, for each risk type, we calculate the mean value of annual disclosure frequency for the given period.

$$M_j = \sum_{t=1}^{T} F_{j,t} / T \tag{3.7}$$

where M_j denotes the mean value of the annual disclosure frequency of risk j $F_{j,t}$ for the sample period T, representing the general condition of foresight on risk j for sample period T.

Then, the FLAI of risk j for period t is defined as

$$FLAI_{j,t} = F_{j,t} / M_j \tag{3.8}$$

$FLAI_{j,t}$ is 1 when $F_{j,t}$ is equal to M_j. Thus, the $FLAI_{j,t}$ of 1 denotes the general condition of foresight on risk j for the sample period T. A $FLAI_{j,t}$ value of greater than 1 denotes that for period t, it is foreseeable that risk j exposed by banks will become more severe than it is under general condition. In contrast, for a $FLAI_{j,t}$ value of less than 1, losses caused by risk j foreseen for period t are considered to become less pronounced than the typical level of general condition.

3.3.3 The Steps to Aggregate Historical and Forward-Looking Financial Statements Data

To aggregate multiple bank risks, some risk aggregation approaches have emerged so far. Simple summation, var-covar and copula approaches are the three main risk aggregation approaches. All of them have strengths and weaknesses (Li et al. 2015). The simple summation approach is one of the most basic and widely used approaches to aggregate risk (Rosenberg and Schuermann 2006; Inanoglu and Jacobs 2009; Kretzschmar et al. 2010). It has several features. One is that it is the briefest one that calculates total risk by adding stand-alone risks. Another is that it is found to be more conservative than other risk aggregation approaches (Embrechts et al. 1999). Such an approach implicitly assumes that all risks are perfectly correlated; that is to say, great losses occur simultaneously, which imposes an upper bound on the actual total risk (Dimakos and Aas 2004). Many papers use the simple summation approach to aggregate marginal risks, such as Rosenberg and Schuermann (2006), Inanoglu and Jacobs (2009) and Kretzschmar et al. (2010).

The simple summation approach assumes that risks are perfectly positively related. The variance–covariance approach only considers the linear correlations between risks by using a matrix of linear correlation coefficients to aggregate individual risks. However, an obvious demerit of the two approaches is that they can only get the total risk VaR but not the entire total risk distribution.

The copula-based approaches use copula functions to specify the required linear and non-linear dependency structures between individual risk distributions (Grundke and Polle 2012). By allowing for a wide variety of dependence between risks, the copula-based approaches are the most promising trend for top-down approaches (Bocker and Hillebrand 2009). However, it is difficult to choose the correct copula functions. Therefore, researchers can adopt the appropriate risk aggregation methods to aggregate marginal risks based on the research needs.

The simple summation approach assumes that all risks are perfectly correlated, i.e. that great losses co-occur, which imposes an upper bound on the true total risk (Dimakos and Aas 2004). Therefore, the simple summation approach is found to be more conservative compared with other risk aggregation approaches (Embrechts et al. 1999). Obviously, the assumption that all risks are perfectly correlated is simple and unrealistic in most cases. Thus, the use of the simple summation approach just allows us to empirically test the impact of forward-looking information on total risk when all risk types occur simultaneously. The total risk value calculated based on historical numerical data recorded in financial statements under the $1 - \alpha$ confidence level for period t is written as:

$$Total\text{-}VaR_t(\alpha) = \sum_j VaR_{j,t}(\alpha) \tag{3.9}$$

where $VaR_{j,t}(\alpha)$ and $Total\text{-}VaR_t(\alpha)$ represent historical marginal risk j and historical total risk value, respectively.

Having obtained historical risks, forward-looking textual risk disclosures are incorporated into bank risk aggregation by using $FLAI_{j,t}$ to adjust historical marginal risks. The forward-looking adjusted total risk value for period t is defined as

$$Adjusted\text{-}Total\text{-}VaR_t(\alpha) = \sum_j FLAI_{j,t} * VaR(\alpha) \tag{3.10}$$

$FLAI_{j,t}$ has an amplifying or reducing effect on historical bank risk $VaR_{j,t}(\alpha)$. Specifically, $Total\text{-}VaR_t(\alpha)$ is a particular case of $Adjusted\text{-}Total\text{-}VaR_t(\alpha)$ when $FLAI_{j,t}$ equals 1, denoting that the general condition of foresight on future bank risk movements is identical to past volatility characteristics and no changes are brought about through the use of forward-looking information. Therefore, besides representing the general condition of foresight on future bank risk movements, $FLAI_{j,t}$ value of equal to 1 is also assumed to represent the historical level of bank risks. Hence, $FLAI_{j,t}$ values greater than 1 increase risk j for period t because, according to forward-looking information, future risk j is considered more severe than historical values. Thus, historical risk values need to be extended to cover future risk losses. $FLAI_{j,t}$ value of less than 1 reduces risk j for period t because forward-looking information projects future risk j to become less severe than historical levels. Hence, the FLAI can increase or reduce historical risk levels based on foresight on risks to attain a more reasonable total risk against future potential losses.

In summary, according to the above steps, we aggregate bank risks based on historical and forward-looking financial statements data. The total risk adjusted by forward-looking textual risk disclosures not only makes the total risk to be determined by historical risks faced by banks but also reflects potential losses that may be incurred by banks in the future. Therefore, the forward-looking adjusted total risk is more reasonable and better able to cover potential total losses that may occur in the future, further guaranteeing the robust operation of banks.

References

Artzner P, Delbaen F, Eber JM, Heath D (1999) Coherent measures of risk. Math Financ 9(3):203–228

Bank for International Settlements Securities and Exchange Commission (SEC) (2005) Securities and exchange commission final rule, release no. 33–8591(FR-75). Retrieved from http://www.sec.gov/rules/final/33-8591.pdf

Bao Y, Datta A (2014) Simultaneously discovering and quantifying risk types from textual risk disclosures. Manage Sci 60(6):1371–1391. https://doi.org/10.1287/mnsc.2014.1930

Basel Committee on Banking Supervision (1988) International convergence of capital measurement and capital standards. Bank for International Settlements, Basel

Basel Committee on Banking Supervision (2006) International convergence of capital measurement and capital standards: a revised framework. Basel, Switzerland

Baselga-Pascual L, Trujillo-Ponce A, Cardone-Riportella C (2015) Factors influencing bank risk in Europe: evidence from the financial crisis. North Am J Econ Finance 34:138–166. https://doi.org/10.1016/j.najef.2015.08.004

Beneish MD, Miller BP, Yohn TL (2015) Macroeconomic evidence on the impact of mandatory IFRS adoption on equity and debt markets. J Account Public Policy 34(1):1–27. https://doi.org/10.1016/j.jaccpubpol.2014.10.002

Bocker K, Hillebrand M (2009) Interaction of market and credit risk: an analysis of inter-risk correlation and risk aggregation. J Risk 11(4):3–29. https://doi.org/10.21314/JOR.2009.198

Breuer T, Jandacka M, Rheinberger K, Summer M (2010) Does adding up of economic capital for market and credit risk amount always to conservative risk estimates? J Bank Finance 34(4):703–712. https://doi.org/10.1016/j.jbankfin.2009.03.013

Campbell JL, Chen HC, Dhaliwal DS, Lu HM, Steele LB (2014) The information content of mandatory risk factor disclosures in corporate filings. Rev Acc Stud 19(1):396–455. https://doi.org/10.1007/s11142-013-9258-3

Dimakos XK, Aas K (2004) Integrated risk modelling. Stat Model 4(4):265–277

Embrechts P, McNeil A, Straumann D (1999) Correlation: pitfalls and alternatives. Risk 12(5):69–71

Grundke P (2010) Top-down approaches for integrated risk management: how accurate are they? European J Oper Res, 203(3), 662–672. https://doi.org/10.1016/j.ejor.2009.09.015

Grundke P, Polle S (2012) Crisis and risk dependencies. Eur J Oper Res 223(2):518–528. https://doi.org/10.1016/j.ejor.2012.06.024

Hartmann P (2010) Interaction of market and credit risk. J Bank Finance 34(4):697–702. https://doi.org/10.1016/j.jbankfin.2009.10.013

Hsu CP, Huang CW, Chiou W (2012) Effectiveness of copula-extreme value theory in estimating value-at-risk: empirical evidence from Asian emerging markets. Rev Quant Financ Acc 39(4):447–468. https://doi.org/10.1007/s11156-011-0261-0

Huang A (2013) Value at risk estimation by quantile regression and kernel estimator. Rev Quant Financ Acc 41(2):225–251

Huang KW, Li ZL (2011) A multilabel text classification algorithm for labeling risk factors in SEC form 10-K. ACM Trans Manage Inf Syst (TMIS) 2(3):1–19. https://doi.org/10.1145/2019618.2019624

Inanoglu H, Jacobs M (2009) Models for risk aggregation and sensitivity analysis: an application to bank economic capital. J Risk Finan Manage 2(1):118–189. https://doi.org/10.3390/jrfm2010118

Jarrow RA, Turnbull SM (2000) The intersection of market and credit risk. J Bank Finance 24(1–2):271–299. https://doi.org/10.1016/S0378-4266(99)00060-6

Jiménez G, Lopez JA, Saurina J (2013) How does competition affect bank risk-taking? J Financ Stab 9(2):185–195. https://doi.org/10.1016/j.jfs.2013.02.004

Kılıç M, Kuzey C (2018) Determinants of forward-looking disclosures in integrated reporting. Manag Audit J 33(1):115–144. https://doi.org/10.1108/MAJ-12-2016-1498

Kretzschmar G, McNeil AJ, Kirchner A (2010) Integrated models of capital adequacy—why banks are undercapitalized. J Bank Finance 34(12):2838–2850. https://doi.org/10.1016/j.jbankfin.2010.02.028

Kupiec PH, Ramirez CD (2013) Bank failures and the cost of systemic risk: evidence from 1900 to 1930. J Finan Intermediation 22(3):285–307. https://doi.org/10.1016/j.jfi.2012.09.005

Li J, Zhu X, Lee CF, Wu D, Feng J, Shi Y (2015) On the aggregation of credit, market and operational risks. Rev Quant Financ Acc 44(1):161–189. https://doi.org/10.1007/s11156-013-0426-0

Liu Y, Ralescu DA (2017) Value-at-risk in uncertain random risk analysis. Inf Sci 391:1–8. https://doi.org/10.1016/j.ins.2017.01.034

Muslu V, Radhakrishnan S, Subramanyam KR, Lim D (2014) Forward-looking MD&A disclosures and the information environment. Manage Sci 61(5):931–948. https://doi.org/10.1287/mnsc.2014.1921

Rosenberg JV, Schuermann T (2006) A general approach to integrated risk management with skewed, fat-tailed risks. J Financ Econ 79(3):569–614. https://doi.org/10.1016/j.jfineco.2005.03.001

Zhu X, Yang SY, Moazeni S (2016) Firm risk identification through topic analysis of textual financial disclosures. Computational Intelligence, IEEE, 1–8. https://doi.org/10.1109/SSCI.2016.7850005

Chapter 4
Bank Risk Aggregation Based on Income Statement

4.1 Introduction

Risk is usually understood in two ways: one is to emphasize the uncertainty of return; the other is to emphasize the uncertainty of loss. According to the narrow definition of risk, risk can only show loss, and there is no possibility of profit from risk. While the broad definition of risk holds that the result of risk may bring loss, profit, or no loss and no profit. Financial risk belongs to broad sense risk.

In the traditional risk measurement, the loss is often taken as the research object, which belongs to the narrow sense of risk. On the one hand, if the loss data cannot be obtained, the risk measurement method using loss data may not be available; on the other hand, this narrow sense risk view which only considers the loss uncertainty is not comprehensive. Mathematically, risk is usually defined as the expected change in the outcome of one or more future events. In particular, financial risk is usually defined as the expected change in the price of an asset or its profit or loss. In banking, risk is usually defined as the fluctuation of profit and loss. Therefore, the measurement of financial risk should consider the broad sense of risk.

The financial statement is an important source to obtain the fluctuation information of profit and loss. Financial statements, also known as external accounting statements, are the accounting statements provided by the accounting entity to reflect the financial status and operation of the accounting entity (Zhu et al. 2018). They generally include balance sheets, income statements, cash flow statements, statements of changes in owner's equity, schedules and notes. The financial statement is a comprehensive reflection of the business performance and financial status of an enterprise in a certain period of time; at the same time, it also reflects the risks that the enterprise undertakes to a certain extent. In financial management, through the analysis of financial statements disclosed by enterprises, we can obtain the risk status of banks to a certain degree (Rosenberg and Schuermann 2006). Therefore, the financial statement is a meaningful way to reveal bank risk information.

J. Li et al., *Financial Statements-Based Bank Risk Aggregation*, Innovation in Risk Analysis, https://doi.org/10.1007/978-981-19-0408-0_4

Like other enterprises, financial statements of financial enterprises contain the information of their operating performance and financial status in a certain period, especially the various risks they bear and the impact of these risks on assets, liabilities, profits and cash flow. Among financial statements, the income statement (or profit and loss statement) directly records the total profit and loss of financial enterprises in a certain period. Moreover, it contains various assets and liabilities affected by different types of risks and the profits and losses of various businesses. Therefore, by analyzing the financial statements of financial enterprises, especially the items in the profit statement, we can obtain information of various types of risks and the overall risks faced by financial enterprises.

According to the risk perception that risk is the uncertainty of profit and loss (Rajan 2006), Kuritzkes and Schuermann (2007) found that establishing the mapping relationship between the income statement items and bank risks can directly obtain first-hand risk profit and loss data. They aggregated risks of major U.S. banks by establishing the mapping relationship between the income statement and different risk types. Due to the significant differences in financial statements between China and the United States, Li et al. (2012) established the mapping relationship between the income statement of Chinese commercial banks and bank risks, integrating the credit, market, liquidity and operational risks of Chinese commercial banks.

Thus, based on the above analysis, collecting risk profit and loss from income statements is a feasible way to collect bank risk data. Specifically, by mapping different bank risk types into income statement items, we can get risk data of different bank risk types. In the empirical analysis, by establishing the mapping relationship between Chinese commercial banks' income statements and credit, market, liquidity, operational, integrated risks, we collect 743 pieces of risk data for each risk type to measure the total risk of the Chinese banking industry.

The rest of this chapter is structured as follows. Section 4.2 illustrates the bank risk aggregation approach based on the income statement. Section 4.3 describes the data used in this chapter. In Sect. 4.4, the approach is applied to aggregating the Chinese banking sector's credit, market, liquidity, and operational risks. Section 4.5 summarizes the conclusions.

4.2 The Income Statement Mapping Approach

4.2.1 The Mapping Relationship Between Risk Types and Income Statement Items

Banking risk is defined in terms of earnings volatility in this research. Therefore, we get risk profit and loss by extracting items from the income statement. Specifically, the corresponding relationship is as Fig. 4.1.

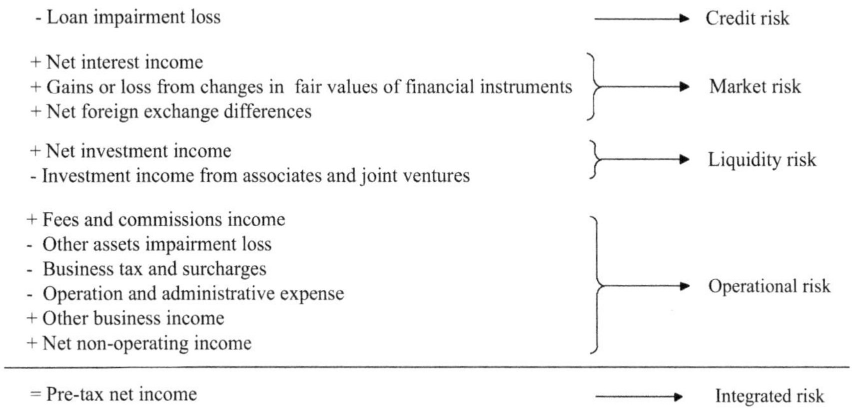

Fig. 4.1 The mapping relationship between risks and income statement

As shown in Fig. 4.1, different items in the income statement can be grouped into different sources of net income, and then mapped into different risk types. In the following mapping of these risk types, some of the income items are mapped into risk types better than others. Overall, the scheme provides a reasonable basis for decomposing earnings volatility into different risk sources.

The cleanest alignment is between credit risk and loan impairment loss. For credit risk, loan provisions, adjusted according to credit risk during a period, are a good proxy because provisions reflect the current credit exposure of banks. However, this proxy still has some disadvantages: defaults are time-lagged and banks can smoothen earnings by controlling loan provisions.

As for market risk, the corresponding items are net interest income, gains or losses from fair values and foreign exchange, and net foreign exchange differences. Variations in net interest income are mainly due to changes in interest rates, i.e. interest rate risk, which is the indispensable part of market risk. Gains or losses from fair values of financial instruments are affected by price fluctuations of financial instruments, and foreign exchange differences are determined by foreign exchange movements.

Liquidity risk in our mapping is caused by the volatility of earnings from investments (gains or losses from investments), which are the primary asset of a bank. However, investment income from associates and joint ventures is generated by long-term equity investment, which is made to control or influence other companies other than to get short-term investment income. Thus, liquidity risk profit and loss equals net investment income minus investment income from associates and joint ventures.

Operational risk, which is the typical and the main non-financial risk, represents the volatility of residual earnings, which cannot be categorized into market, credit or liquidity risk. It is hard to identify the sources of operational risk from the income statement. The remaining income items: fees and commissions income, other assets impairment loss, business tax and surcharges, operation and administrative expense, other business income and net non-operating income serve as a proxy for operational risk.

In the last line, pre-tax net income can be mapped into the integrated risk, which reflects the overall operating conditions of banks over a period.

4.2.2 Procedure of Risk Measurement and Aggregation

We measure risks based on the definition of risks in terms of earnings volatility. Therefore, after establishing the relationship between income statement items and risk types, the risk profit and loss need to be converted into a return-based measure for direct comparison across banks. The approach in our analysis is to divide earnings by risk weighted assets to yield a return on risk weighted assets (RORWA), which is preferable to un-weighted total assets since it makes some adjustment for the risk of the underlying assets. For neutralizing bank effects, we need to compute deviations in RORWA by subtracting from the average RORWA for each bank and the mean-adjusted RORWA reflect the bank risk over a period essentially.

Let $Y_{i,j,t}$ be the earnings for bank i, risk j in period t, and let $RWA_{i,t}$ be the corresponding level of risk-weighted assets. We then define RORWA for the ith bank, j risk in period t as

$$r_{i,j,t} = \frac{Y_{i,j,t}}{RWA_{i,t}} \tag{4.1}$$

and mean-adjusted RORWA as

$$\Delta_{i,j,t} = \left(\frac{Y_{i,j,t}}{RWA_{i,t}}\right) - \left(\frac{1}{T_i}\sum_{t=1}^{T_i} r_{i,j,t}\right) \tag{4.2}$$

where bank i is observed for T_i periods. The mean-adjusted RORWA is the risk deviation from expected risk return.

Finally, we obtain a typical bank's risk returns to model marginal risk distributions. For a typical bank, its risk return is determined by market information and individual information. The market information is composed by all sample banks' deviation from risk return. Thus, by combining all sample banks' deviation from risk return and the typical bank's $(i = k)$ expected risk return, we finally compute a typical bank's risk return. Specifically, the typical bank's risk return is written as

$$r_{k,j,t} = \overline{r}_{k,j} + \Delta_{j,t} = \overline{r}_{k,j} + \sum_i \Delta_{i,j,t} \tag{4.3}$$

where $r_{k,j,t}$ is the risk return of bank k, risk j in period t. $\overline{r}_{k,j}$ stands for the expected risk return of bank k and risk j. $\Delta_{j,t}$ denotes the summation of deviation from risk return of risk j in period t of all sample banks, reflecting the market information of risk j in period t. For simplicity, we refer to (Eq. 4.3) as RORWA in the following context.

After getting RORWA, we employ VaR as the basic tool to measure the risk. The risk in this book is defined in terms of earnings volatility, and VaR is a probabilistic method of measuring the potential loss or earnings volatility in portfolio value over a given time and for a given distribution. Generally, VaR has become a standard model for measuring and assessing risk. There are three ways to measure VaR: historical simulation method, parameter method and Monte Carlo simulation method. In this chapter, we will employ the historical simulation method to measure risks of Chinese banks, of which the core is establishing risk return distribution by using historical samples.

4.3 Data Description

Since only listed banks' financial reports are publicly available and new accounting standards were applied in 2007, we collected quarterly panel data over 2007–2018 from early 16 A-share listed Chinese commercial banks (Table 4.1) to ensure the consistency of accounts. The quarterly data of ABC and CEB from 2007 to 2009 are unavailable because they were listed in 2010. Besides, 2007-Q2 data of BOBJ, 2007-Q1 data of BONJ, BONB and CCB are also missing. Getting rid of these exceptional cases, we finally obtained 743 pieces of valid data to model individual risk distributions.

Our empirical analysis is based on quarterly data, while loan impairment loss and loan loss provision are disclosed only in annual and semi-annual financial reports. Hence, we need to make simple assumptions to obtain quarterly data of these

Table 4.1 The sample of all 16 early-listed commercial banks of China

No	Bank	No	Bank
1	Industrial and Commercial Bank of China (ICBC)	9	Huaxia Bank (HXB)
2	China Construction Bank (CCB)	10	Industrial Bank (IB)
3	Bank of China (BOC)	11	China Everbright Bank (CEB)
4	Agriculture Bank of China (ABC)	12	China Minsheng Bank (CMB)
5	Bank of Communications (BOCOM)	13	China CITIC Bank (CITIC)
6	China Merchants Bank (CMB)	14	Bank of Beijing (BOBJ)
7	Pingan Bank (PAB)	15	Bank of Ningbo (BONB)
8	SPD Bank (SPDB)	16	Bank of Nanjing (BONJ)

accounts. Specifically, loan impairment loss, which is part of assets impairment loss, can be calculated based on known quarterly assets impairment loss. Specifically, we first calculate R, which is a ratio of loan impairment loss to assets impairment loss based on annual and semi-annual data. Then, we calculate the mean value of this ratio over the sample period ($\overline{R}$).

Herein, we make a simple assumption that the quarterly R is equal to $\overline{R}$. Thus, the quarterly loan impairment loss is obtained by multiplying $\overline{R}$ and quarterly assets impairment loss. Likewise, the quarterly loan loss provision, which is determined by the quality of loans, can be obtained based on the known quarterly loans. Specifically, we define R' as the ratio of loan loss provision to loans and $\overline{R}'$ as the mean value of R' over the sample period. Therefore, the quarterly loan loss provision is obtained by multiplying $\overline{R}'$ with quarterly loans based on the assumption that the quarterly R' is equal to $\overline{R}'$.

Among the sample of all 16 early-listed Chinese commercial banks from 2007 to 2018, the number of financial statements data for a single bank is up to 48, which is too small to perform the empirical analysis. Thus, in order to address the problem of data shortage and provide empirical insights into the total risk of Chinese commercial banks, we construct hypothetical banks following Rosenberg and Schuermann (2006), Kretzschmar et al. (2010) and Alessandri and Drehmann (2010). In particular, we use median assets to characterize hypothetical banks, and specifically, a large hypothetical bank, a small hypothetical bank and a hypothetical general bank are constructed for comparison.

"The big four" state-owned banks are the four largest banks by assets in the Chinese banking system. However, ABC went public relatively late so that the amount of financial statements data is relatively smaller. Therefore, the asset size of the large hypothetical bank is the average of the rest three state-owned banks (ICBC, BOC and CCB). Correspondingly, the asset size of the small hypothetical bank is the average of the three smallest banks by assets (BOBJ, BONB and BONJ). By using this median approach, we construct three hypothetical banks that differ in scale. At the end of 2018, the bank sizes in terms of risk-weighted assets for the large, general and small hypothetical banks are 14,564, 6641 and 1161 billion CNY, respectively.

For either of these three hypothetical banks, the amount of data is 743, which is much larger than that of a real-world bank. Furthermore, the hypothetical banks constructed by us capture the characteristics of real-world banks' asset sizes, so they are the typical banks in the Chinese banking system. In a word, performing empirical analysis based on typical hypothetical banks not only addresses the problem of data shortage but also achieves general conclusions.

4.4 Empirical Results

4.4.1 Total Risk Results of Chinese Commercial Bank

For the three hypothetical banks, the marginal risk distribution is decided by the deviation from risk return and the expected risk return by referring to Eq. (4.3). The deviation from risk return, which reflects the macroeconomic background and operating conditions of the banking industry, decides the shape of marginal risk distribution. The expected risk return that reflects a bank's features decides the horizontal axis coordinates of the marginal risk distribution. The characteristics of deviation from risk return are presented numerically in Table 4.2 and the shapes of marginal risk distributions are visually shown in Fig. 4.2.

Specifically, in terms of mean-adjusted RORWA, which is the risk deviation, the market risk has the highest volatility (11.72%) and fattest tails (Kurtosis = 165.83). The volatility of credit risk comes second with a value of 2.05%. The volatility values of operational and liquidity risks are 0.99% and 0.25%, respectively. The kurtosis of liquidity comes second (17.01). The negative kurtosis presents in credit risk (−0.39) and the relatively lower positive kurtosis for operational risk (1.98) suggest that credit and operational risks have thinner tails, with volatilities of them being 0.50% and 0.31%, respectively. The shape of credit risk distribution is nearly symmetric, with the value of skewness being 0.53. However, liquidity risk is moderately right-skewed at 3.18 and market risk is more significantly right-skewed at 10.08. The operational risk is left-skewed at −0.88.

Table 4.3 records VaR values of individual risks at different confidence levels. The larger negative value or smaller positive value of VaR corresponds to the higher level of risk. At a 99.9% confidence level, the credit risk increased with the decrease of bank size. The VaR values of credit risk for the large, general and small banks are 0.79%, 0.37% and 0.36%, respectively. Regarding the market risk, only the VaR value of the large bank is positive (0.60%), indicating that the large bank has a better market risk management ability. However, for the general and small banks, the VaR values are negative, which are −6.05% and −4.12%, respectively. As for the liquidity risk, the VaR values for the large, general and small banks are negative, which are −1.01%, −1.01% and −0.21%, respectively, indicating that the small bank has the better liquidity risk management ability. For the operational risk, the VaR values for the large, general and small banks are negative, which are −3.77%, −5.93% and −3.98%, respectively, indicating that the large bank has the better liquidity risk

Table 4.2 Descriptive statistics for four marginal risks of three hypothetical banks

Statistic	Credit risk	Market risk	Liquidity risk	Operational risk
σ (%)	2.05	11.72	0.25	0.99
Skewness	0.54	10.08	3.18	−0.88
Kurtosis	−0.39	165.83	17.01	1.98

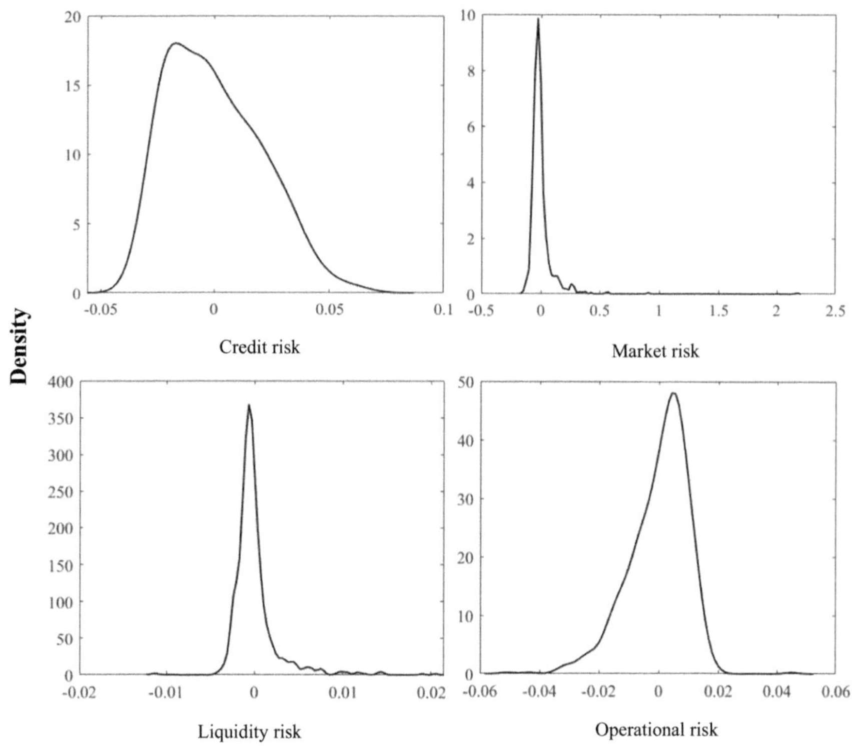

Fig. 4.2 Marginal distributions of credit, market, liquidity and operational risks

management ability. Furthermore, for the general bank, the most considerable risk is market risk, then is the operational risk, followed by liquidity risk, and the lowest risk is credit risk. However, for the large bank, the most significant risk is operational risk. Then is the liquidity risk. The third serious risk is market risk, and the lowest risk is credit risk.

After getting VaR values of marginal risks, we can then get the total risk distribution by mapping it into the pre-tax net income of income statement. The distribution of total risk is shown in Fig. 4.3. The statistics of the total risk distribution are recorded in Table 4.4. Specifically, the volatility is 0.99% and the kurtosis is 0.23. The total risk is moderately left-skewed at −0.40.

Our quarterly financial statements data enable a quarterly view of total risk while the typical horizon of losses is one year. To transform the total quarterly loss into the annual total loss, we apply the square-root-of-time rule which is commonly used to scale an estimated quantile of a return distribution to a lower frequency T by the multiplication of (Danielsson and Zigrand 2006). Table 4.5 gives a summary overview of the three hypothetical banks' total risks in 2018.

As shown in Table 4.5, at 99.9% confidence level, the total risk VaR values are 0.82%, 0.70% and 0.36% for the large, general and small banks, respectively. We can

Table 4.3 Four marginal risks VaR of three hypothetical banks/%

		Credit risk (%)	Market risk (%)	Liquidity risk (%)	Operational risk (%)
Large bank	0.1th percentile	0.79	0.60	−1.01	−3.77
	1st percentile	0.91	1.49	−0.24	−3.41
	2nd percentile	0.93	1.75	−0.20	−3.32
	5th percentile	1.38	2.29	−0.12	−2.12
General bank	0.1th percentile	0.37	−6.05	−1.01	−5.93
	1st percentile	0.81	−4.46	−0.20	−4.08
	2nd percentile	0.97	−1.92	−0.16	−3.58
	5th percentile	1.27	0.21	−0.12	−3.01
Small bank	0.1th percentile	0.36	−4.12	−0.21	−3.98
	1st percentile	0.80	−3.45	−0.18	−3.41
	2nd percentile	0.88	−3.29	−0.14	−3.39
	5th percentile	1.04	−2.53	−0.12	−2.60

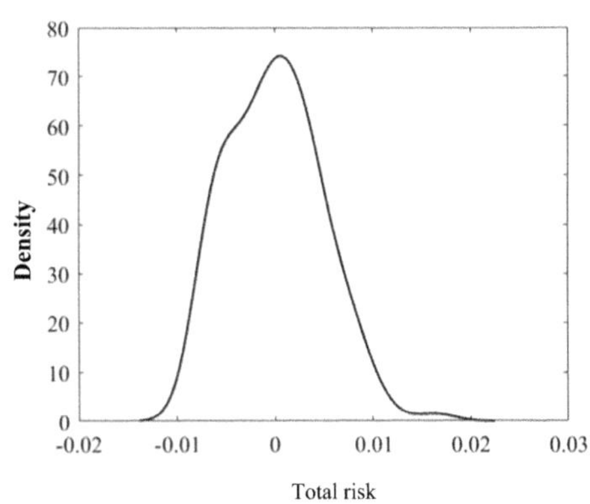

Fig. 4.3 The distribution of total risk based on pre-tax net income

Table 4.4 Descriptive statistics for total risk of three hypothetical banks

Statistic	σ (%)	Kurtosis	Skewness
Kurtosis	0.99	0.23	−0.40

Table 4.5 2018 total risk VaR and total losses of the three hypothetical banks

Confidence level (%)	99.9	99	98	95
Large hypothetical bank				
Total risk (VaR) (%)	0.82	0.83	0.86	0.69
General hypothetical bank				
Total risk (VaR) (%)	0.70	0.72	0.74	0.81
Small hypothetical bank				
Total risk (VaR) (%)	0.36	0.80	0.88	1.04

see that the total risk VaR values declined with the decrease of bank size, indicating that the total risk increases with the decrease of bank size. Therefore, the large bank has the lowest total risk while the small bank has the highest total risk.

4.4.2 Risk Diversification Results

The total risks obtained by mapping total risk into pre-tax net income of income statement are recorded in Table 4.5. Here we aim to analyze the risk diversification benefit of the bank risk aggregation approach based on income statement, which measures total risk by mapping total risk into pre-tax net income of income statement. As per Eq. (4.4), the diversification benefit is defined as the reduction in total risk VaR due to less-than-perfect correlation across individual risks (Begley et al. 2016).

$$\delta = \frac{\sum_{i=1}^{d} VaR_i - (total\ VaR)}{\sum_{i=1}^{d} VaR_i} \tag{4.4}$$

where δ denotes the diversification coefficient, d denotes the number of individual risks, and VaR_i denotes the VaR of individual risk i.

To aggregate different risks into total risk, the simple summation approach is one of the most fundamental risk aggregation approaches and has been widely used in many studies (Rosenberg and Schuermann 2006; Inanoglu and Jacobs 2009; Kretzschmar et al. 2010; Li et al. 2018). Simple summation, variance–covariance and copula approaches are three commonly used risk aggregation approaches (Li et al. 2015). However, the variance–covariance and copula approaches are not applicable in this chapter. Specifically, the variance–covariance and copula approach aggregate multiple risks by modeling the inter-risk correlations based on one-to-one data of different risks. However, by processing data using Eqs. (4.1) to (4.3), the obtained data of credit, market and operational risks for the typical bank are not corresponding one by one. Thus, it is not practical to compute correlations between different risks based on the risk data of the typical bank. That is the reason why the variance–covariance and copula risk aggregation approaches are not applicable in this chapter.

Table 4.6 The diversification benefits of different risk aggregation approaches

Confidence level (%)	99.9	99	98	95
Large hypothetical bank				
Total risk (VaR) (%)	0.82	0.83	0.86	0.96
Simple summation (%)	−3.39	−1.25	−0.84	1.43
Diversification effect (%)	124.19	166.40	202.38	32.87
General hypothetical bank				
Total risk (VaR) (%)	0.55	0.68	0.76	0.93
Simple summation (%)	−12.62	−7.93	−4.69	−1.65
Diversification effect (%)	104.36	108.58	116.20	156.36
Small hypothetical bank				
Total risk (VaR) (%)	0.70	0.72	0.74	0.81
Simple summation (%)	−7.95	−6.24	−5.94	−4.21
Diversification effect (%)	108.81	111.54	112.46	119.24

The simple summation approach assumes that all risks are perfectly correlated, i.e. that great losses coincide, which imposes an upper bound on the true total risk (Dimakos and Aas 2004). Thus, the simple summation approach is found to be more conservative compared with other risk aggregation approaches (Embrechts et al. 1999). Obviously, the assumption that all risks are perfectly correlated is simple and unrealistic in most cases. Thus, using the simple summation approach only allows us to empirically analyze the total risk when all risk types occur simultaneously. The total risk value calculated based on historical numerical data recorded in financial statements under the $1 - \alpha$ confidence level for period t is written as:

$$Total\text{-}VaR_t(\alpha) = \sum_j VaR_{j,t}(\alpha) \tag{4.5}$$

where $VaR_{j,t}(\alpha)$ and $Total\text{-}VaR_t(\alpha)$ represent historical marginal risk j and historical total risk value, respectively.

The diversification coefficients of three banks at different confidence levels are recorded in Table 4.6. Specifically, at 99.9% confidence level, the diversification coefficients of the large, general and small banks are 124.19%, 104.36% and 108.81%, respectively.

4.5 Conclusions

In this chapter, we describe the bank risk aggregation approach based on the income statement. Specifically, mapping different risk types into different items recorded in income statements allows us to get risk data for each risk. Furthermore, by mapping pre-tax net income into total risk, we obtain the total risk distribution. In the empirical analysis, we apply this approach to aggregate credit, market, liquidity and operational risks by using a sample of all 16 early-listed Chinese listed commercial banks for 2007–2018. Then we empirically study the difference of bank risks in different scales by constructing three large, medium and small hypothetical Chinese commercial banks.

Our empirical results show that at 99.9% confidence level, the total risk VaR values are 0.82%, 0.70% and 0.36% for the large, general and small banks, respectively. It is clear that the total risk VaR values declined with the decrease of bank size, indicating that the total risk increases with the decrease of bank size. Thus, the large bank has the lowest total risk while the small bank has the highest total risk.

However, this chapter has several limitations. The correspondence between risk types and financial statements is kind of rough. For example, net interest income bears credit risk and market risk simultaneously. However, the item of net interest income is mapped into credit risk only. In future studies, the employment of other information may help calibrate the corresponding relationship to some extent.

References

Alessandri P, Drehmann M (2010) An economic capital model integrating credit and interest rate risk in the banking book. J Bank Finance 34(4):730–742. https://doi.org/10.1016/j.jbankfin.2009.06.012

Begley TA, Purnanandam AK, Zheng K (2016) The strategic under-reporting of bank risk. Rev Finan Stud 30(10):3376–3415. https://doi.org/10.1093/rfs/hhx036

Danielsson J, Zigrand JP (2006) On time-scaling of risk and the square-root-of-time rule. J Bank Finance 30(10):2701–2713. https://doi.org/10.1016/j.jbankfin.2005.10.002

Dimakos XK, Aas K (2004) Integrated risk modelling. Stat Model 4(4):265–277. https://doi.org/10.1191/1471082X04st079oa

Embrechts P, McNeil A, Straumann D (1999) Correlation: pitfalls and alternatives. Risk 12(5):69–71

Inanoglu H, Jacobs M (2009) Models for risk aggregation and sensitivity analysis: an application to bank economic capital. J Risk Finan Manage 2(1):118–189. https://doi.org/10.3390/jrfm2010118

Kretzschmar G, McNeil AJ, Kirchner A (2010) Integrated models of capital adequacy-why banks are undercapitalized. J Bank Finance 34(12):2838–2850. https://doi.org/10.1016/j.jbankfin.2010.02.028

Kuritzkes A, Schuermann T (2007) What we know, don't know and can't know about bank risk: A view from the trenches. In: Diebod FX et al (eds) The known, the unknown and the unknowable in financial risk management. Princeton University Press, Princeton

Li J, Wei L, Lee CF, Zhu X, Wu D (2018) Financial statements based bank risk aggregation. Rev Quant Financ Acc 50(3):673–694. https://doi.org/10.1007/s11156-017-0642-0

Li J, Yi S, Zhu X, Feng J (2012) Mutual information based copulas to aggregate banking risks. In: 2012 fifth international conference on business intelligence and financial engineering, Lanzhou

Li J, Zhu X, Lee CF, Wu D, Feng J, Shi Y (2015) On the aggregation of credit, market and operational risks. Rev Quant Financ Acc 44(1):161–189. https://doi.org/10.1007/s11156-013-0426-0

Rajan RG (2006) Has finance made the world riskier? Eur Financ Manag 12(4):499–533

Rosenberg JV, Schuermann TA (2006) General approach to integrated risk management with skewed, fat-tailed risks. J Financ Econ 79(3):569–614. https://doi.org/10.1016/j.jfineco.2005.03.001

Zhu X, Wei L, Wu D, Li J (2018) A general framework for constructing bank risk data sets. J Risk 21(1):37–59. https://doi.org/10.21314/JOR.2018.393

Chapter 5
A "Factor-Integral" Approach to Solve the Low-Frequency Problem of Income Statement Data

5.1 Introduction

Credit and market risks are the main risks faced by banks (Bellini 2013; Li et al. 2018). However, they have often been treated as unrelated sources of risk: the risk types have been measured separately, managed separately, and economic capital (EC) against each risk type has been assessed separately in early studies (Breuer et al. 2010; Adrian 2017). As a matter of fact, market and credit risks are intrinsically related to each other and are not separable (Jarrow and Turnbull 2000; Drehmann et al. 2010). Thus, risk aggregation approaches that incorporate dependent risks into a single metric by considering complex correlations between them are proposed (Imbierowicz and Rauch 2014). The risk aggregation approaches can quantify the integrated loss arising from dependent risks, which are crucial to arrive at the accurate integrated EC, especially as proven by the subprime crisis (Bocker and Hillebrand 2009; Dionne 2013).

Two kinds of risk aggregation approaches have thus far emerged, including top-down approaches (TDAs) and bottom-up approaches (BUAs) (Hartmann 2010; Grundke 2013). TDAs separately measure each risk type and then aggregate individual risks by modeling the complex correlations between them through the commonly used simple summation, variance–covariance and copula-based approaches (Rosenberg and Schuermann 2006; Aas et al. 2007). However, it is difficult to sufficiently capture the complex interactions between various risk types, which may lead to sizable biases in overall risk estimates (Grundke 2010; Bellini 2013).

In contrast, the other alternative way to aggregate different risks is to follow the bottom-up approach (BUA). In the BUA, the different risk types are correlated through their underlying risk factors. Thus, BUAs model the complex interactions between different risk types by considering the potential interactions of their underlying risk factors (Brockmann and Kalkbrener 2010). The multivariate dependence between risk factors and the influence of risk factors on the different risk types allows for a more accurate determination of the integrated risk distribution (Grundke 2010).

J. Li et al., *Financial Statements-Based Bank Risk Aggregation*, Innovation in Risk Analysis, https://doi.org/10.1007/978-981-19-0408-0_5

Therefore, BUAs are assumed to be more accurate than TDAs (Basel Committee on Banking Supervision, BCBS for short 2009).

However, a major obstacle faced by both the existing two kinds of risk aggregation approaches is the problem of data shortage. Since severe risk events occur infrequently, scarce historical data are available (Aas et al. 2007; Zhu et al. 2018). Instead, recent researches have collected risk data from financial statements to solve the problem of data shortage. However, the risk data recorded in financial statements are low-frequency (usually published quarterly). Especially for an immature financial market like China, with only 25 listed commercial banks and a very short listing period. So the problem of risk data shortage is even worse in this situation (Zhu et al. 2018). Adding or omitting several data might substantially influence the distribution of bank risk and thus result in a bias in integrated risk estimation (Rosenberg and Schuermann 2006; Li et al. 2012). Thus, the data shortage is one of the major challenges involved in risk aggregation (Zhu et al. 2018). Many risk aggregation approaches typically fail due to the low quantity of data available in the banks (BCBS 2010; Bignozzi and Tsanakas 2014).

Many researchers have found that risk loss is caused by changes in risk factors (Wei et al. 2019). Specifically, credit and market risk losses are driven by some risk factors, such as interest rate, equity index, credit spread and exchange rate (Breuer et al. 2010; Grundke 2010). Although risk data is low-frequency, their underlying risk factors are high-frequency, as high as daily (Grundke 2009). Thus, it is an alternative way to address the problem of data shortage in risk aggregation by explaining the low-frequency risk data through a much richer history of high-frequency risk factor data (Rosenberg and Schuermann 2006).

Therefore, this chapter proposes a new factor-integral approach to obtain integrated loss due to credit and market risks. It can transform the aggregation of low-frequency risk data into the integral of high-frequency risk factor data, which can address the problem of data shortage in risk aggregation and obtain a more accurate and stable integrated loss distribution. Specifically, credit and market risk data are explained by common risk factors and their respective idiosyncratic risk factors. The existence of common risk factors makes credit and market risks naturally correlated. Thus, the integrated loss due to credit and market risks is caused by changes in their common risk factors and idiosyncratic risk factors. By integrating the common factors and idiosyncratic factors, we obtain integrated loss due to credit and market risks. Through a dataset covering 16 Chinese listed commercial banks spanning the period 2007–2019, our proposed approach is empirically illustrated and compared with risk aggregation based on low-frequency risk data by aggregating the credit risk and market risk of the Chinese banking system.

The rest of this chapter is structured as follows. Section 5.2 illustrates the proposed factor-integral approach and its specific implementation steps. Section 5.3 describes the data used in this chapter. In Sect. 5.4, the proposed approach is applied to aggregating the credit and market risks of the Chinese banking sector. Section 5.5 summarizes the conclusions.

5.2 The Factor-Integral Approach

5.2.1 The Concept of the Factor-Integral Approach

Broadly, risk aggregation approaches are proposed to incorporate dependent risks into a single metric by considering complex correlations between them (Rosenberg and Schuermann 2006). This chapter proposes a new risk aggregation approach called the factor-integral approach to obtain integrated loss due to credit and market risks.

The concept of the proposed approach is illustrated in Fig. 5.1. Specifically, credit risk is the loss resulting from the failure of obligors to honor their payments (Li et al. 2015). Market risk measures the loss associated with a bank's trading positions arising from adverse movements in market factors such as asset prices, foreign exchange rates, or interest rates (BCBS 2006; Bocker and Hillebrand 2009). The risk data for credit risk and market risk are low-frequency. However, researchers have found that credit risk and market risk are affected by many risk factors. The data of risk factors are high-frequency (as high as daily).

Among these risk factors, some simultaneously affect credit risk and market risk. The existence of common risk factors between credit and market risks made them naturally correlated. Besides the common risk factors, the residual risk factors are the respective idiosyncratic factors for credit risk and market risk, reflecting the respective characteristics of credit risk and market risk. Thus, credit risk and market risk are explained by common risk factors and their respective idiosyncratic risk factor. The integrated loss due to credit and market risks is determined by changes in common risk factors and idiosyncratic risk factors of credit and market risks. By integrating the high-frequency common risk factors and idiosyncratic risk factors of

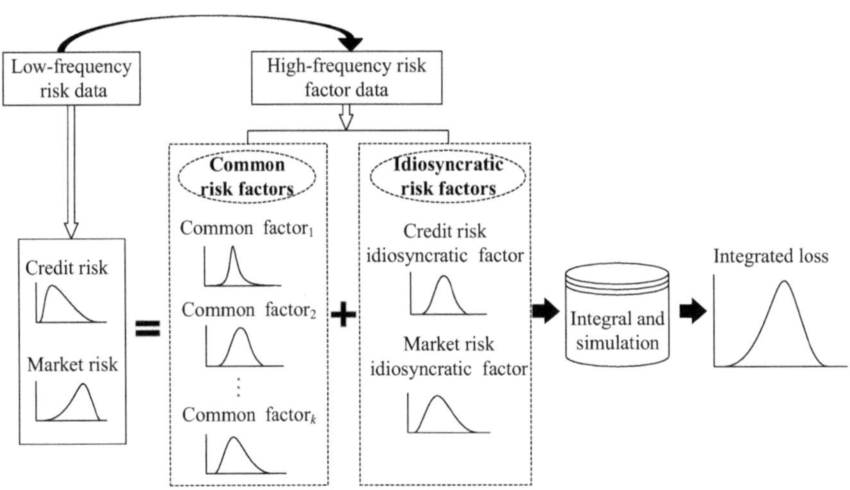

Fig. 5.1 The concept of the proposed factor-integral approach

credit and market risks, the integrated loss due to credit risk and market risk can be obtained. By doing this, the aggregation of low-frequency risk data is transformed into the integral of high-frequency risk factor data.

The mathematical illustration of this proposed factor-integral approach is illustrated in detail as follows. Firstly, we identify notations in our proposed factor-integral approach.

$\mathbf{X}$ = k×1 vector of common risk factors
$\boldsymbol{\beta}^{\mathbf{c}}$ = k×1 vector of common risk factor sensitivities for credit risk
$\boldsymbol{\beta}^{\mathbf{m}}$ = k×1 vector of common risk factor sensitivities for market risk
ε^{c} = residual credit risk that is not explained by the common risk factors
ε^{m} = residual market risk that is not explained by the common risk factors.

Here, we assume that credit risk and market risk are affected by k common factors $X_1, X_2, \ldots, X_k$. Let L^c denote credit risk, and L^m denote market risk. Let $\beta_1^c, \beta_2^c, \ldots, \beta_k^c$ and $\beta_1^m, \beta_2^m, \ldots, \beta_k^m$ denote the parameters of common factors. Let ε^c denote the idiosyncratic factor of credit risk, and ε^m denote the idiosyncratic factor of market risk. Then, the functional relationships between risk loss and risk factors are shown by (Eq. 5.1) and (Eq. 5.2).

$$L^c = F_c(X_1, X_2, \ldots, X_k) + \varepsilon^c \tag{5.1}$$

$$L^m = F_m(X_1, X_2, \ldots, X_k) + \varepsilon^m \tag{5.2}$$

Our aim is to obtain integrated loss due to credit and market risks. The integrated loss distribution is constructed based on the definition of cumulative probability density distribution. From a mathematical perspective, the cumulative probability density distribution describes the probability of a random variable falling on an interval, which is obtained by integrating the probability density function of the continuous random variable. As discussed above, the integrated loss in this chapter is defined as loss due to credit and market risks, which is determined by changes in common risk factors and idiosyncratic risk factors of credit and market risks (Pianosi and Wagener 2015). So let L denote the integrated loss; the definition of the cumulative probability density distribution of integrated loss is written as (Eq. 5.3). Then combining (Eq. 5.1) and (Eq. 5.2), (Eq. 5.4) to (Eq. 5.6) can be derived.

$$P(L \leq r) \tag{5.3}$$

$$= P\left(L^c + L^m \leq r\right) \tag{5.4}$$

$$= P\left[F_c(X_1, X_2, \ldots, X_k) + F_m(X_1, X_2, \ldots, X_k) + \varepsilon^c + \varepsilon^m \leq r\right] \tag{5.5}$$

$$= \iint \cdots \int_D f\left(\varepsilon^c\right) d\varepsilon^c f\left(\varepsilon^m\right) d\varepsilon^m f(X_1) dX_1 \ldots f(X_k) dX_k \tag{5.6}$$

where the integral region D is that $X_1, X_2, \ldots, X_k$ satisfied a condition like

$$F_c(X_1, X_2, \ldots, X_k) + F_m(X_1, X_2, \ldots, X_k) + \varepsilon^c + \varepsilon^m \leq r \tag{5.7}$$

From (Eq. 5.3) to (Eq. 5.6), it is clear to see that the integrated loss distribution is constructed by integrating the probability density functions of the common risk factors and idiosyncratic risk factors of credit and market risks. Once some integrated loss values are given, the corresponding probability can be calculated through multiple integrals of risk factors. A large number of integrated loss values and their corresponding probabilities will construct the empirical cumulative probability density distribution of integrated loss.

Overall, by using (Eq. 5.1) to (Eq. 5.6), we transform the aggregation of low-frequency risk data into the integral of high-frequency risk factor data based on the mathematical calculation of cumulative probability density distribution. This is the reason why we name our proposed risk aggregation approach the factor-integral approach. The risk aggregation results are believed to be more accurate and robust than estimations directly from the low-frequency risk data. In practical application, the common risk factors are discrete variables. Based on plenty of Monte Carlo simulations of common risk factors, we can simulate integrated loss with L^c and L^m defined as in (Eq. 5.1) and (Eq. 5.2). By doing this, an empirical cumulative probability density function for the integrated loss L can be generated.

Our proposed factor-integral approach depends on two key components. One is the selection of common risk factors. If common factors are found to explain most of the risk uncertainty, the analysis will have been instrumental. Suppose they explain only a small fraction of total risk uncertainty. In that case, the residual risk will have to be examined to see whether many more common factors exist that are worth investigating and including in the analysis (Alexander and Pezier 2003; Breuer et al. 2010). The other is to find the best fitting distributions of common risk factors. The integral result completely depends on the distributions of risk factors. The best-fitting distributions for common risk factors can reflect the movements in common risk factors more precisely and eventually lead to an accurate result of the integrated loss.

5.2.1.1 A Simple Linear Factor Model

A simple solution to explain the risk is through a linear regression model. Taking credit risk and market risk as the dependent variables and the common factors as the independent variables, the parameters of common risk factors can be easily calculated by two linear regressions. In this case, a linear factor-integral model is given as follows:

$$\begin{aligned} L^c &= \alpha^c + \boldsymbol{\beta}^{\mathbf{c}\prime}\mathbf{X} + \varepsilon^c \\ &= \alpha^c + \beta_1^c X_1 + \beta_2^c X_2 + \cdots + \beta_k^c X_k + \varepsilon^c \end{aligned} \tag{5.8}$$

$$\begin{aligned} L^m &= \alpha^m + \boldsymbol{\beta}^{\mathbf{m}'}\mathbf{X} + \varepsilon^m \\ &= \alpha^m + \beta_1^m X_1 + \beta_2^m X_2 + \cdots + \beta_k^m X_k + \varepsilon^m \end{aligned} \tag{5.9}$$

$$\begin{aligned} P(L \le r) &= P\left(L^c + L^m \le r\right) \\ &= P\left[\left(\beta_1^c + \beta_1^m\right)X_1 + \left(\beta_2^c + \beta_2^m\right)X_2 + \cdots \right. \\ &\quad \left. + \left(\beta_k^c + \beta_k^m\right)X_k + \alpha^c + \alpha^m + \varepsilon^c + \varepsilon^m \le r\right] \\ &= P\left[\left(\beta_1^c + \beta_1^m\right)X_1 + \left(\beta_2^c + \beta_2^m\right)X_2 + \cdots \right. \\ &\quad \left. + \left(\beta_k^c + \beta_k^m\right)X_k + \varepsilon^c + \varepsilon^m \le r - \alpha^c - \alpha^m\right] \\ &= \iint \cdots \int_{D_1} f\left(\varepsilon^c\right)d\varepsilon^c f\left(\varepsilon^m\right)d\varepsilon^m f(X_1)dX_1 \ldots f(X_k)dX_k \end{aligned} \tag{5.10}$$

where the integral region D_1 is that X_1, X_2, ..., X_k satisfied a condition like $\left(\beta_1^c + \beta_1^m\right)X_1 + \left(\beta_2^c + \beta_2^m\right)X_2 + \cdots + \left(\beta_k^c + \beta_k^m\right)X_k + \varepsilon^c + \varepsilon^m \le r - \alpha^c - \alpha^m$.

Specifically, credit risk L^c and market risk L^m are linear functions of k common factors X_1, X_2,..., X_k. Then, the linear relationships between risk and risk factors are established by (Eq. 5.8) and (Eq. 5.9). By applying (Eq. 5.8) and (Eq. 5.9) to the general (Eq. 5.6), we finally obtain (Eq. 5.10), which derives the cumulative probability of linear integrated loss L through a $k + 2$ numerical integral.

5.2.1.2 A Non-linear Factor Model

It may be inappropriate to assume that risk is linearly related to the risk factors, as in (Eq. 5.8) and (Eq. 5.9). In most forms, we have a non-linear factor-integral model instead. Alexander and Pezier (2003) deemed that an extreme variation that only occurred in a factor with very small sensitivity would not induce an extreme risk variation (Alexander and Pezier 2003). In addition, higher-order models have higher computation costs and may cause an over-fitting problem (Dunis et al. 2011). In this case, a second-order approximation is given by:

$$L^c = \alpha^c + \boldsymbol{\beta}^{\mathbf{c}'}\mathbf{X} + 1/2\mathbf{X}'\boldsymbol{\gamma}^{\mathbf{c}}\mathbf{X} + \varepsilon^c \tag{5.11}$$

$$L^m = \alpha^m + \boldsymbol{\beta}^{m'}\mathbf{X} + 1/2\mathbf{X}'\boldsymbol{\gamma}^{m}\mathbf{X} + \varepsilon^m \tag{5.12}$$

$$\begin{aligned} & P(L \le r) \\ & = P\left(L^c + L^m \le r\right) \\ & = P\left[\sum_{i=1}^{k}\left(\beta_i^c + \beta_i^m\right)X_i + 1/2\left[\sum_{i=1}^{k}\left(\gamma_{ii}^c + \gamma_{ii}^m\right)X_i^2 + 2\sum_{i,j} r_{i,j}X_iX_j\right]\right. \\ & \quad \left. + \alpha^c + \alpha^m + \varepsilon^c + \varepsilon^m \le r\right] \end{aligned}$$

$$
\begin{aligned}
&= P\left[\sum_{i=1}^{k}\left(\beta_i^c+\beta_i^m\right)X_i + 1/2\left[\sum_{i=1}^{k}\left(\gamma_{ii}^c+\gamma_{ii}^m\right)X_i^2 + 2\sum_{i,j}^{k} r_{i,j}X_iX_j\right]\right.\\
&\quad \left. +\varepsilon^c+\varepsilon^m \le r-\alpha^c-\alpha^m\right]\\
&= \iint\cdots\int_{D_2} f\left(\varepsilon^c\right)d\varepsilon^c f\left(\varepsilon^m\right)d\varepsilon^m f(X_1)dX_1\ldots f(X_k)dX_k\\
&\quad i=1,2,\ldots,k;\ j=1,2,\ldots,k;\ i\neq j
\end{aligned}
\tag{5.13}
$$

where the integral region D_2 is that X_1, X_2, ..., X_k satisfied a condition like $\sum_{i=1}^{k}\left(\beta_i^c+\beta_i^m\right)X_i + 1/2\left[\sum_{i=1}^{k}\left(\gamma_{ii}^c+\gamma_{ii}^m\right)X_i^2 + 2\sum_{i,j}^{k} r_{i,j}X_iX_j\right] + \varepsilon^c + \varepsilon^m \le r-\alpha^c-\alpha^m$.

To be specific, credit risk L^c (Eq. 5.11) and market risk L^m (Eq. 5.12) are deduced by a second-order Tailor expansion of k common factors $X_1, X_2, \ldots, X_k$. By applying (Eq. 5.11) and (Eq. 5.12) into the general (Eq. 5.6), we finally obtain (Eq. 5.13), which is used to calculate the cumulative probability of non-linear integrated loss L through a $k+2$ numerical integral.

5.2.2 *Specific Steps of Using the Factor-Integral Approach*

In this section, the detailed implementation steps of the proposed factor-integral approach are given. At the outset, credit and market risk data and common factors to credit and market risks need to be determined. Then, the proposed factor-integral approach can be employed to aggregate credit and market risks. As discussed above in Sect. 5.2.1, the integrated loss due to credit and market risks is determined by changes in the common risk factors and idiosyncratic risk factors. So following the five steps illustrated in Fig. 5.2, we can obtain the integrated loss distribution using the proposed factor-integral approach. The specific procedure of using the proposed factor-integral approach is described in detail as follows.

Input: Data of L^c, L^m and $X_1, X_2, \ldots, X_k$
Step 1 Parameter estimation

Firstly, we need to estimate the parameters of functions $L^c = F_c(X_1, X_2, \ldots, X_k) + \varepsilon^c$ and $L^m = F_m(X_1, X_2, \ldots, X_k) + \varepsilon^m$. Since risk data are low-frequency, while common risk factors data are high-frequency, in this step, the common risk factor data need to be adjusted to the same frequency as the risk data.

Step 2 Idiosyncratic risk factor calculation

After the parameters of functions are estimated, the idiosyncratic risk factor data ε^c and ε^m can be calculated by using (Eq. 5.14) and (Eq. 5.15). The idiosyncratic risk factor here is the residual of the regression.

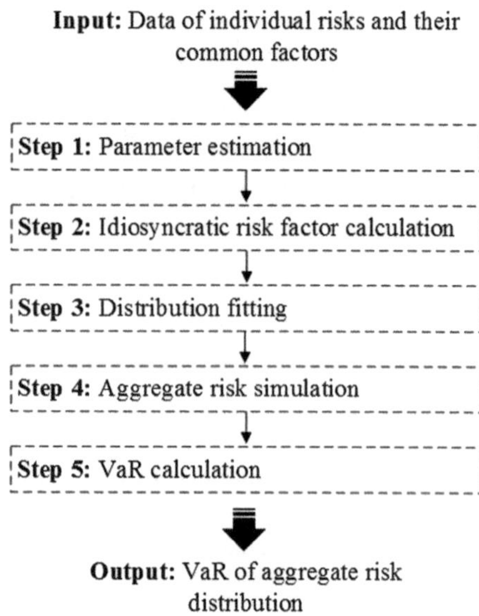

Fig. 5.2 The procedure of the factor-integral approach

$$\varepsilon^c = L^c - F_c(X_1, X_2, \ldots, X_k) \tag{5.14}$$

$$\varepsilon^m = L^m - F_m(X_1, X_2, \ldots, X_k) \tag{5.15}$$

Step 3 Distribution fitting

Based on (Eq. 5.3) to (Eq. 5.6), the distribution of integrated loss is constructed by integrating the probability density functions of common risk factors and idiosyncratic risk factors of credit and market risks. So in this step, we find the best-fitting distributions $f(X_1)$, $f(X_2)$, …, $f(X_k)$ for common risk factors, $f(\varepsilon^c)$ and $f(\varepsilon^m)$ for idiosyncratic risk factors.

The common risk factors $X_1, X_2, \ldots, X_k$ that we use here are the original collected high-frequency data without adjustment because high-frequency data are considered more accurate and stable when fitting a distribution. The goodness-of-fit test is adopted to examine whether a theoretical distribution is a fit for an empirical distribution. Among many goodness-of-fit tests, the Kolmogorov–Smirnov (KS) and chi-square tests are widely used (Zhu et al. 2019).

KS test is based on the maximal discrepancy between the expected and the observed cumulative distribution (Bernal et al. 2014). The null hypothesis (H_0) is that the empirical distribution of observed data conforms to a known theoretical distribution. The KS test statistic D_{KS}, which denotes the maximum discrepancy, is written as:

$$D_{KS} = \sup_x |F_n(x) - F(x)| \tag{5.16}$$

where $\sup_x$ denotes the supremum, $F_n(x)$ and $F(x)$ represent the empirical cumulative distribution function and the theoretical cumulative distribution function, respectively. Once the sample size n and the significance level α are determined, we can obtain the threshold of the maximum discrepancy. If the calculated D_{KS} is smaller than the threshold, we accept the null hypothesis that the empirical distribution of observed data conforms to the known theoretical distribution. The KS test provides a p-value. The significance level is generally set at 5%. The larger the p-value, the better the effect of the fitting is. Therefore, if the p-value is more significant than 0.05, we accept the null hypothesis and thus find the proper distributions for risk factors.

The Chi-square test is a traditional measure for evaluating goodness-of-fit. The principle of chi-square is that the distribution obeyed by a random variable is deduced based on the empirical distribution of its sample observations. The chi-square test assesses the magnitude of the discrepancy between the empirical distribution of the sample and the assumed theoretical distribution (Nye and Drasgow 2011). If the magnitude of the discrepancy is small enough, the empirical distribution of the sample is considered to be consistent with the assumed theoretical distribution. The null hypothesis (H_0) assumes that the random variable obeys a theoretical distribution. Dividing the sample data into n groups, the chi-square test statistic $\chi^2$is defined as:

$$\chi^2 = \sum_{i=1}^{n} \frac{(f_i - F_i)^2}{F_i} \tag{5.17}$$

where f_i and F_i represent the actual and theoretical number of data in group i, respectively. When n or f_i is large enough, the chi-square test statistic χ^2 approaches the chi-square distribution. From (Eq. 5.17), it is clear to see that the smaller the value of χ^2, the closer the empirical distribution and the assumed theoretical distribution is. The chi-square test provides a p-value computed using the chi-square distribution with degrees-of-freedom $n - 1 - k$, denoted $\chi^2(n - 1 - k)$. k is the number of parameters to be estimated in the assumed theoretically hypothetical distribution. The acceptable threshold level of chi-square test is the p-value greater than 0.05.

Thus, in this step, we adopted the KS test to find the proper distribution of all risk factors. In addition, the maximum likelihood method is used to estimate the parameters of the factor distributions. As a robustness check, an alternative goodness-of-fit chi-square test is also adopted.

Step 4 Integrated loss simulation

After the distributions of all factors $f(X_1), f(X_2), \ldots, f(X_k), f(\varepsilon^c)$ and $f(\varepsilon^m)$ are fitted, as shown in (Eq. 5.6), the credit risk and market risk can be aggregated by integrating over these risk factor distributions. To attain the distribution of integrated loss, we need to conduct a certain number of simulations. In each simulation, given an r value, its corresponding probability is calculated by (Eq. 5.6).

When the number of simulations reaches a certain threshold, this simulated empirical distribution is close to the actual distribution. The larger the number of simulations is, the closer the simulated integrated loss distribution will be to the actual distribution while the longer computational time is required. To balance simulation accuracy and time cost, it is generally accepted that simulation times N should conform to (Eq. 5.18).

$$N \times \alpha \times (1 - \alpha) \geq 50 \tag{5.18}$$

where α is the confidence level. For risk management purposes, confidence level α is generally set larger than 90%. In this chapter, we calculate value-at-risk (VaR) at confidence levels of 95%, 98%, 99% and 99.9%; thus, we set N as 100,000, which meets the requirement of (Eq. 5.18).

Step 5 VaR and EC calculation

Pioneered by J.P. Morgan, VaR has become a standard measure used in financial risk measurement (Kabaila and Mainzer 2018). Therefore, we also employ VaR to measure the integrated loss based on the simulated integrated loss distribution (Step 5). VaR at a specific confidence level $\alpha \in (0, 1)$ is defined as the smallest number l such that the probability of loss L exceeding l is not larger than $(1 - \alpha)$:

$$VaR = \inf\{l : P(L \leq l) \leq (1 - \alpha)\} \tag{5.19}$$

Once we have obtained the integrated loss distribution, we can further calculate the EC. EC is defined as the capital that a financial institution requires against unexpected risk losses, which is the difference between the expected loss and integrated loss (Furman et al. 2017). The EC computed via our proposed approach are the necessary amount of capital needed to absorbing potential losses due to the credit and market risks.

Output: VaR and EC of the integrated loss distribution

The data of risk and risk factors can be inputted, and then VaR of the integrated loss and EC can be outputted by the above five steps of the proposed factor-integral approach. In the end, it is noteworthy that the aggregation of credit and market risks is illustrated here. Theoretically, this approach can easily be extended to aggregate other risk types or more than two types of risks, as long as these risks have common risk factors.

5.3 Data Description

5.3.1 *Low-Frequency Risk Data*

Since profit and loss (P&L) arising from bank risks within specific periods are recorded and summarized in the income statement, looking for risk proxies from income statements provides a reasonable basis for decomposing profits and losses into risk sources. Some researchers suggested that mapping income statement P&L items into risk types is an alternative way to collect risk data (Rosenberg and Schuermann 2006; Li et al. 2018). Thus, in this chapter, the risk data of credit and market risks are collected from P&L items recorded in income statements of Chinese commercial banks.

Previous studies took provisions as the risk proxy of credit risk (Kuritzkes and Schuermann 2007; Zhu et al. 2018). The income statement item of provisions, which is a charge for incurred loan losses arising from credit default, is inherently related to the definition of credit risk. Inanoglu and Jacobs (2009) mapped credit risk into charge-offs. Based on a robustness check made by Kuritzkes and Schuermann (2007), the choice between provision and charge-offs appears to make little difference to measure credit risk. Therefore, although provisions do not comprehensively represent credit risk, mapping provisions into credit risk is an acceptable choice according to previous studies. As for the market risk, taking trading income as the risk proxy has reached a consensus (Rosenberg and Schuermann 2006; Kuritzkes and Schuermann 2007; Inanoglu and Jacobs 2009).

In China, however, trading income is not specified in the income statement. Li et al. (2018) suggested that the sum of changes in fair value gains or losses, net investment income, and net foreign exchange differences are the proxies for market risk in the Chinese banking sector as these three items are essentially determined by market factors. Besides, since the item of trading income records P&L related to the trading activities, previous studies took the trading income as the market risk proxy to measure the market risk (Rosenberg and Schuermann 2006; Kuritzkes and Schuermann 2007). In this chapter, we also map market risk into trading income.

Although there are now 25 A-share listed Chinese commercial banks, 9 of them were listed too late in 2016 or 2017. As discussed in Sect. 4.3 of Chap. 4, we collected P&L data from the 16 early A-share listed Chinese commercial banks (Table 4.1) over 2007–2019. Getting rid of missing data, we finally obtained 807 pieces of valid data for each credit and market risks.

Risk P&L items from income statements are not comparable among different banks as banks differ in scale, capital allocation, investment strategy, and management level (Rosenberg and Schuermann 2006). To allow for a direct comparison across banks, risk P&L needs to be converted into a "risk-return" based measure. Following Kuritzkes and Schuermann (2007), dividing risk P&L by risk-weighted assets (RWA) can yield a return on RWA measure. The specific risk proxies are shown in Table 5.1.

Table 5.1 The proxies of credit risk and market risk

Risk type	Proxy
Credit risk	– Provisions/RWA
Market risk	(Net investment income + net foreign exchange differences + the changes in fair value gains or losses)/RWA

5.3.2 High-Frequency Risk Factor Data

Having determined risk proxies for credit and market risks, we need to identify risk factors that commonly affect credit and market risks. Some previous studies have made contributions to identifying common risk factors for credit and market risks. Specifically, Jarrow and Turnbull (2000) and Breuer et al. (2010) regarded interest rate and equity market index as common factors for credit and market risks. Besides interest rate and equity index, Alexander and Peizer (2003) also pointed out that credit spread is a common risk factor, while Aas et al. (2007) suggested that exchange rate is another common risk factor. Kretzschmar et al. (2010) used these four types of common risk factors, including interest rate, equity index, credit spread and exchange rate simultaneously. Besides the above four types of common risk factors, commodity index was regarded as a common risk factor by Medova and Smith (2005).

The potential risk common factors identified by Grundke (2010) also included macroeconomic factors (e.g. gross domestic product, thereafter GDP). In summary, our literature review indicates that the following six types of common factors, i.e. interest rate, equity index, credit spread, exchange rate, commodity index and macroeconomic factors, are widely accepted as common risk factors in aggregating credit and market risks. Therefore, here we comprehensively select these six types of common risk factors in our proposed factor-integral approach, and their detailed descriptions are shown in Table 5.2.

Specifically, the Shanghai Stock Exchange (SSE) index and the CSI300 index are two widely used stock indexes in China designed to replicate the equity market's performance. The 1-year AAA government bond yield (R1) and the 10-year AAA government bond yield (R10) are used to represent interest rates. The Sino-US exchange rate (e) is selected for the exchange rate because the US dollar is still the most important foreign currency for China. Another variable is the Cninfo RMB Currency index (CI), which describes the overall movement of the RMB against a basket of major currencies. The AA and BBB + credit spreads are calculated by the AA and BBB + corporate bond yield minus the risk-free government bond yield, which represent high-quality and low-quality enterprise credit, respectively. The Commodity Research Bureau (CRB) commodity index, which is the world's oldest leading commodity index, is designed to isolate and reveal the directional movement of prices in overall commodity trades. Producer price index (PPI), consumer price index (CPI) and gross domestic product (GDP) are usually regarded as major

Table 5.2 Common risk factors and their descriptions

Number	Factor type	Specific variable	Description
1	Equity market index	Shanghai Stock Exchange index (SSE)	Ln-difference of SSE index
2		CSI300 index	Ln-difference of CSI300 index
3	Interest rate	R1	Ln difference of 1-year AAA government bond yield
4		R10	Ln-difference of 10-year AAA government bond yield
5	Exchange rate	The Sino-US exchange rate (e)	Ln-difference of the Sino-US exchange rate
6		Cninfo RMB Currency index (CI)	Ln-difference of the Cninfo RMB Currency index
7	Credit spread	AA spread (AA)	Ln-difference of AA spread
8		BBB + spread (BBB +)	Ln-difference of BBB + spread
9	Commodity index	Commodity Research Bureau (CRB) commodity index	Ln-difference of CRB commodity index
10	Macroeconomics	Producer price index (PPI)	Ln-difference of PPI growth rate
11		Consumer price index (CPI)	Ln-difference of CPI growth rate
12		Gross domestic product (GDP)	Ln-difference of GDP growth rate

variables to reflect the macroeconomic environment. Excepting PPI, CPI and GDP, which are quarterly data, the other 9 risk factor variables are daily data.

All data on common risk factors are from the Wind database (http://www.wind.com.cn/). The data period is from 2007 to 2019.

In addition to common risk factors themselves, their volatilities also affect risks (Alexander and Peizer 2003; Rosenberg and Schuermann 2006). Therefore, we employ the generalized autoregressive conditional heteroskedasticity (GARCH) (1, 1) approach, a widely used and classical method that has worked well in most applied situations to capture volatility features of common risk factors (Wei et al. 2010). To ensure the stationarity required for the GARCH analysis, as described in Table 5.3, the measures of common risk factors are the ln-differences from the original data series.

Table 5.3 Results of the pooled regression for credit risk

Variable	Coefficient	Std. error	t-statistic	Prob
C	−0.004649	0.000476	−9.762320	0.0000
GDP	0.006838	0.001019	6.710234	0.0000
CRB	0.018763	0.002811	6.674002	0.0000
E	0.048582	0.011258	4.315265	0.0000
var-e	12.441940	3.410526	−3.648099	0.0003
var-SSE	−0.106578	0.019676	−5.416580	0.0000
var-GDP	0.073550	0.023022	3.194735	0.0015
var-csi300var-AA	−0.308117	0.034841	−8.843593	0.0000
var-PPI	6.586520	1.226566	5.369884	0.0000
var-BBB +	1.405461	0.278545	5.045725	0.0000
PPI	−0.042268	0.008662	−4.879725	0.0000
var-CSI300	0.097876	0.021446	4.563921	0.0000
CI	0.033576	0.007294	4.603418	0.0000
R-squared	0.7678			

5.4 Empirical Results

5.4.1 Aggregate Result of Credit and Market Risk

5.4.1.1 Linear Integrated Results

In the regression, daily risk factors and their volatilities have to be transformed into quarterly data by summing data within a quarter to match the frequency of the credit and market risk returns derived from quarterly financial statements (Rosenberg and Schuermann 2006). After obtaining quarterly risk factors and their volatilities, the factors' sensitivities to the risk types can be estimated based on panel risk-return data collected from 16 Chinese listed commercial banks over the sample period from 2007 Q1 to 2019 Q4. Since we choose common risk factors comprehensively, we adopt the stepwise regression to keep significant risk factors.

Under the simple linear model of the factor-integral approach, the risk returns of market and credit risks are the explained variables, while the risk factors (12 common risk factors) and their volatilities (12 volatilities) are the explanatory variables in the stepwise linear pooled regressions. Therefore, the regression coefficients are the factors' sensitivities to risk returns.

Table 5.3 presents sensitivities for the credit risk stepwise linear pooled regression. It is clear to see that the 12 significant variables at the 5% level explained 76.78% of the credit risk returns, and the other 12 variables that are not statistically significant are omitted from the final model. We can see that GDP and its volatility, the CRB commodity index, the Sino-US exchange rate, the Cninfo RMB currency index and the volatilities of PPI, the BBB + credit spread and the CSI300 equity index are positively correlated with the credit risk return. Meanwhile, the credit risk return has a negative relationship with the volatilities of the Sino-US exchange rate, the SSE equity index, the AA credit spread and PPI.

For the market risk, the stepwise linear pooled regression result shows that 17 variables are not significant at the 5% level; thus, they are omitted from the final model. The remaining seven risk factors and implied volatilities explained 63.74% of the market risk returns, as shown in Table 5.4. It is found that the BBB + credit spread, GDP and its volatilities and the volatility of the Cninfo RMB currency index are negatively correlated with the market risk return. The market risk return has a positive relationship with the Sino-US exchange rate, the 1-year government bond yield R1 and the volatility of the AA credit spread.

After the parameters of common factors are estimated, the idiosyncratic factors of credit risk and market risk, ε^c and ε^m, can be calculated by (Eq. 5.14) and (Eq. 5.15). However, we obtain panel data of idiosyncratic factors as we use panel data to regress. To fit distributions of idiosyncratic factors, the residuals of panel regression are averaged among 16 banks to get time series data of idiosyncratic factors, which offers a general reflection of idiosyncratic factors in the Chinese banking sector. Since common risk factors have significant explanatory power for risk P&L, this simple mean value processing for idiosyncratic factors affects risk P&L little.

In the separable credit and market stepwise linear pooled regression models, there are a total of 15 variables of common risk factors affecting credit risk and market risk. To examine which distribution of these 15 variables and two idiosyncratic factors follow, as discussed in Step 3 in Sect. 5.2.2, we adopt the KS test and chi-square test. Besides, the original collected high-frequency risk factor data are used to fit distribution because high-frequency data are considered more accurate and stable when fitting a distribution.

The p-values of the KS test results and chi-square test results are more significant than 0.05, which means that we found properly fitting distributions for the historical high-frequency data of risk factors. Figure 5.3 shows that these 15 significant

Table 5.4 Results of the pooled regression for market risk

Variable	Coefficient	Std. error	t-statistic	Prob
C	0.001064	8.62e−5	12.34	0.0000
BBB+	−0.001711	0.000354	−4.84	0.0000
E	0.002888	0.001439	2.01	0.0454
GDP	−0.001202	0.000189	−6.35	0.0000
var_GDP	−0.023056	0.004612	−5.00	0.0000
var_CI	−0.319351	0.123940	−2.58	0.0103
var_AA	0.016107	0.006363	2.53	0.0117
R1	0.000210	8.20e−5	2.56	0.0107
R-squared	0.6374			

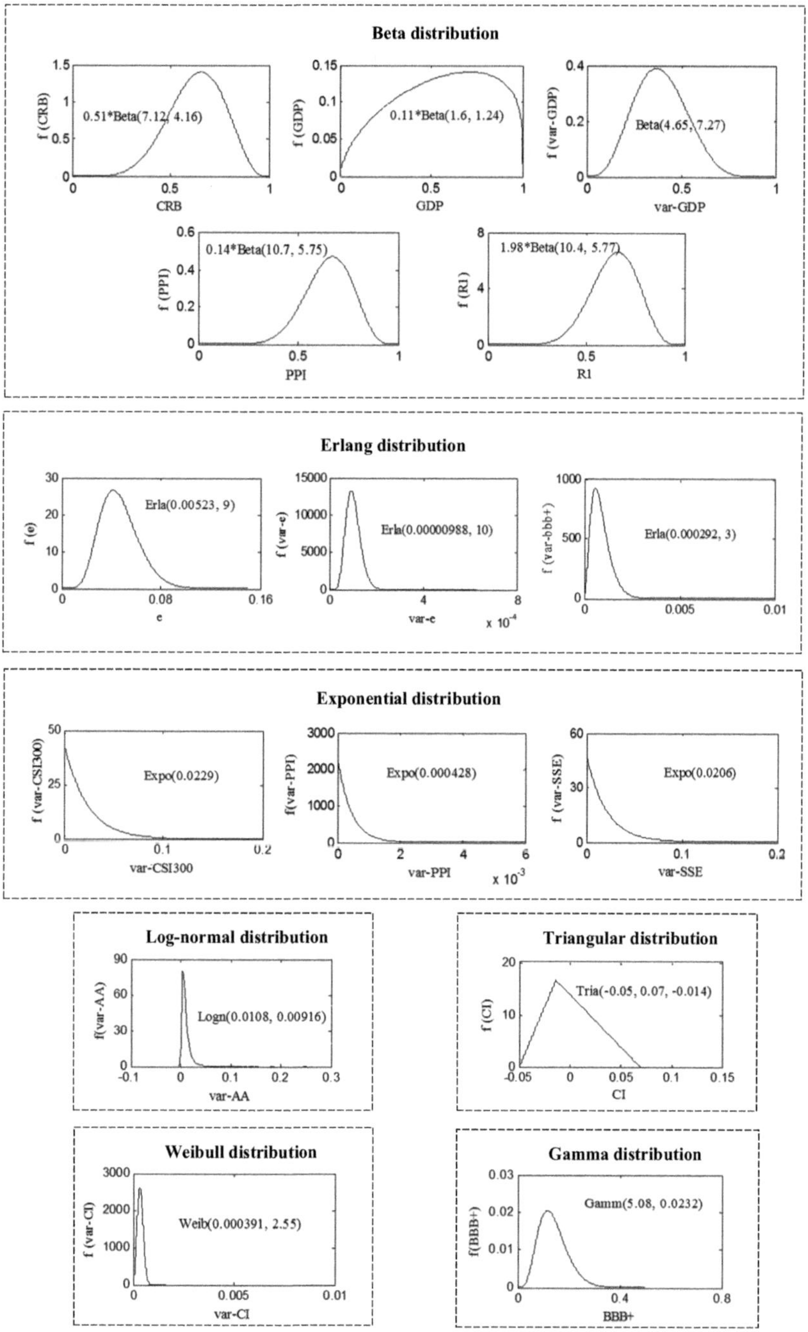

Fig. 5.3 The fitted distributions of common risk factors and their volatilities

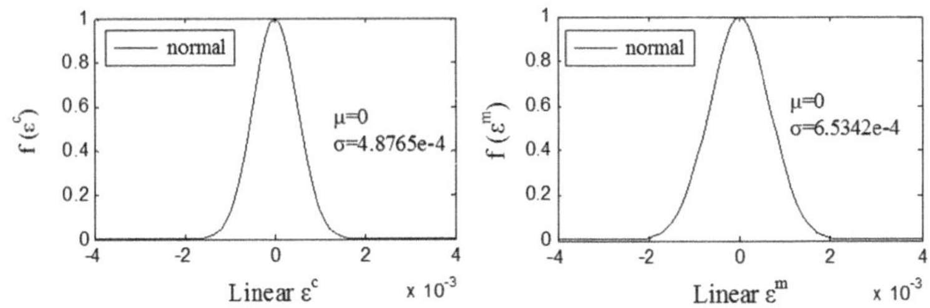

Fig. 5.4 The fitted normal distributions of idiosyncratic risk factors in the linear factor-integral model

common factors and their volatilities follow seven different types of distribution, and Fig. 5.4 illustrates that the two idiosyncratic factors follow normal distributions with different fitted parameters. Specifically, the CRB commodity index, GDP and its volatility, PPI and the 1-year government bond yield R1 follow beta distributions with different fitted parameters. The Sino-US exchange rate and its volatility and the volatility of the BBB + credit spread follow Erlang distributions with different fitted parameters. The volatilities of CSI300, PPI and SSE follow exponential distributions with different fitted parameters. The volatility of the AA credit spread is log-normally distributed. The Cninfo RMB Currency index CI follows a triangular distribution, while its volatility follows a Weibull distribution. For the BBB + credit spread, it follows a gamma distribution.

Having determined the distributions of common and idiosyncratic risk factors, we simulate high-frequency data for risk factors and then transform them into quarterly data by summing data within a quarter to match the frequency of the credit and market risk returns derived from quarterly financial statements (Rosenberg and Schuermann 2006). According to Step 4 in Sect. 5.2.2, we then calculate 100,000 integrated loss values and their corresponding probabilities to construct the empirical cumulative integrated loss distribution. Based on the left tail of the distribution, the VaR of integrated loss can be easily calculated.

Because we use quarterly data in regression functions, the time horizon for the calculated VaR is a quarter. However, the Regulatory Committee requires that the time horizon for the capital charge should be set as one year. A simple but widely used way of scaling is the *square-root-of-time rule*, which is used to scale an estimated quantile of a return distribution to a lower frequency T by the multiplication of $\sqrt{T}$ (Danielsson and Zigrand, 2006). Using this method, the quarterly VaR is multiplied by the square root of 4 (4 quarters in a year) to scale the quarterly VaR to the final VaR with a time horizon of one year. In 2019, the RWA of the 16 listed Chinese banks was 105,694 billion CNY. The 2019 annual integrated VaR and EC with different confidence levels are recorded in Table 5.5. For example, at the 99.9% confidence level, the annual integrated VaR is -5.12%. The 2019 annual EC that Chinese banking should keep to absorb potential unexpected loss arising from credit and market risks is 5411 billion CNY.

Table 5.5 2019 annual integrated risk of the Chinese banking system under the linear factor-integral model

Confidence level (%)	95	98	99	99.9
VaR	−2.22	−3.00	−3.58	−5.12
EC (billion CNY)	2724.26	3368.18	3849.99	5411

5.4.1.2 Non-linear Integrated Results

In addition to the simple linear model under the factor-integral approach applied to the Chinese banking sector, we also aggregate the Chinese banking credit and market risks through a non-linear model of the factor-integral approach introduced in Sect. 5.2.1.2. To keep significant variables, we adopt stepwise quadratic pooled regression. The stepwise quadratic regression results of credit and market risks are recorded in Tables 5.6 and 5.7, respectively.

Quadratic regression is a function of linear and quadratic terms (including square terms and cross terms). Compared with linear regression, in which 12 significant single terms explained 76.78% of the credit risk (see Table 5.4), there are additional

Table 5.6 Results of the quadratic pooled regression for credit risk

Variable	Coefficient	Std. error	t-statistic	Prob
C	−0.003997	0.000455	−8.791541	0.0000
GDP	0.010330	0.001543	6.693871	0.0000
CRB	0.038715	0.004052	9.554380	0.0000
E	0.180082	0.014206	12.67613	0.0000
var-e	−14.90621	2.934710	−5.079276	0.0000
var-SSE	−0.086979	0.017686	−4.917936	0.0000
var-GDP	−0.152545	0.028527	−5.347290	0.0000
var-AA	−0.553347	0.041832	13.22770	0.0000
var-PPI	10.120480	1.157802	8.741116	0.0000
var-BBB +	4.242168	0.568138	7.466794	0.0000
PPI	−0.047191	0.009717	4.856510	0.0000
var-CSI300	0.316345	0.034933	9.055794	0.0000
CI	0.041121	0.007473	5.502490	0.0000
CRBvar-SSE	−1.678268	0.263565	6.367644	0.0000
GDPvar-AA	−1.027509	0.194818	−5.274193	0.0000
evar-SSE	−11.18930	1.181864	−9.467496	0.0000
PPIe	−1.844520	0.319793	−5.767638	0.0000
CRBCI	0.258457	0.064286	4.020451	0.0001
var-BBB + var-CSI300	−202.1316	46.13014	−4.381769	0.0000
R-squared	0.8755			

Table 5.7 Results of the pooled quadratic regression for market risk

Variable	Coefficient	Std. error	t-statistic	Prob
C	0.001226	9.51e−5	12.89323	0.0000
BBB +	−0.008242	0.001763	−4.676339	0.0000
E	0.004002	0.001565	2.556975	0.0109
GDP	−0.001334	0.000192	−6.963878	0.0000
var-GDP	−0.02994	0.005142	−5.822932	0.0000
var-CI	−0.548673	0.135657	−4.044549	0.0001
var-AA	0.01421	0.006563	2.165141	0.0310
R1	0.00027	8.50e−5	3.178327	0.0016
BBB + var-CI	8.901852	3.230233	2.755792	0.0061
BBB + var-GDP	0.280504	0.095914	2.924533	0.0037
eR1	0.011895	0.005773	2.060636	0.0400
R-squared	0.7681			

six significant cross-quadratic terms at the 5% level in the non-linear quadratic regression, explaining 87.55% of the credit risk (Table 5.6). Compared with linear market risk regression, the additional three significant quadratic terms at the 5% level make the R-squared of market risk quadratic regression increase from 63.74 to 76.81% (Table 5.7).

After the coefficients of common factors are estimated, the idiosyncratic factors of credit risk and market risk, ε^c and ε^m in the nonlinear model, can also be calculated by (Eq. 5.14) and (Eq. 5.15). As discussed in Sect. 5.4.1.1, the average residuals are used to fit distributions of idiosyncratic factors. By adopting the KS test and chi-square test, these two idiosyncratic factors are found to follow normal distributions with different fitted parameters (Fig. 5.5).

After the fitted distributions of common and idiosyncratic risk factors have been found, we simulate high-frequency risk factor data and then transform them into quarterly data by summing data within a quarter to match the frequency of the credit and market risk returns. Based on Step 4 in Sect. 5.2.2, 100,000 integrated risk values and their corresponding probabilities are calculated to construct the empirical

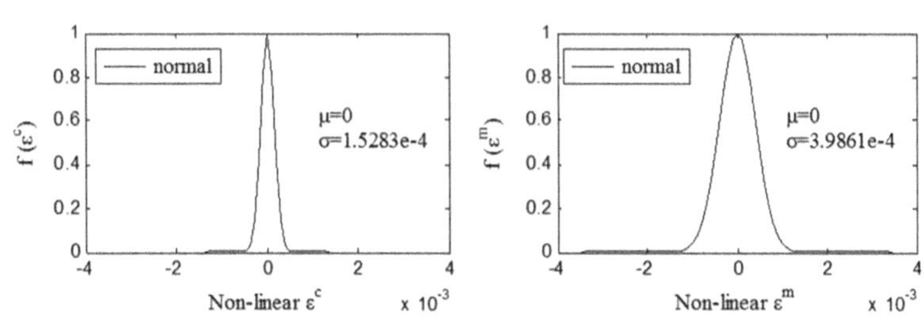

Fig. 5.5 The fitted distributions of idiosyncratic risk factors in the non-linear factor-integral model

Table 5.8 2019 annual integrated risk of Chinese banking system under the non-linear factor-integral approach

Confidence level (%)	95	98	99	99.9
VaR	−1.91	−2.76	−3.28	−4.93
EC (billion CNY)	2019	2917	3467	5210

cumulative integrated loss distribution. The corresponding 2019 annual integrated VaR and EC of the Chinese banking system obtained through the non-linear factor-integral approach are recorded in Table 5.8. At the 99.9% confidence level, the integrated VaR derived from the non-linear factor-integral approach is −4.93%. In 2019, the Chinese banking system should have kept 5210 billion CNY to protect against the potential unexpected losses arising from credit and market risks.

5.4.2 Aggregation Results Comparisons

In Sects. 5.4.1.1 and 5.4.1.2, we obtain the integrated VaR and EC of the Chinese banking system with linear and non-linear factor-integral models, respectively. Theoretically, the integrated results derived from our proposed factor-integral approach are more accurate and robust based on a much richer history of risk factor movements than P&L data that we have observed at a quarterly frequency (Rosenberg and Schuermann 2006). To empirically illustrate the difference, we adopt the following steps to aggregate risks based on low-frequency risk P&L data for comparison. We label this approach as the low-frequency-risk data approach.

Step 1: Input time series of risk P&L data for credit risk and market risk, denoted by quarterly L_{ct} and L_{mt}, respectively.
Step 2: Let L_t denote integrated loss due to credit and market risks. Let $L_t = L_{ct} + L_{mt}$, and directly fit the distribution of integrated loss L_t with its low-frequency historical risk P&L data collected from income statement; the distribution is denoted by $f(L_t)$.
Step 3: Let the integrated loss distribution be $f(L_t)$, and calculate the quarterly VaR of integrated loss. Finally, by scaling the quarterly integrated VaR to annual VaR, the EC of the Chinese banking sector is obtained.

Thus, according to our summarized mapping relationship in Table 5.1, we obtain 807 quarterly risk P&L data each for credit and market risks over 2007–2019 to calculate integrated risk P&L.

The characteristics of integrated loss distributions under the linear and non-linear factor-integral approaches and the low-frequency-risk data approach are presented numerically in Table 5.9. The shapes of these three integrated loss distributions are visually shown in Fig. 5.6.

Table 5.9 Descriptive statistics for integrated loss distributions obtained via the factor-integral approach and the low-frequency-risk data approach

Integrated loss distribution	Proposed factor-integral approach		Low-frequency-risk data approach
	Linear	Non-linear	
σ	0.0103	0.0089	0.0042
U	0.0054	0.0062	−0.0030
Skewness	0.0759	0.0536	−1.5278
Kurtosis	0.0886	0.0611	5.8231

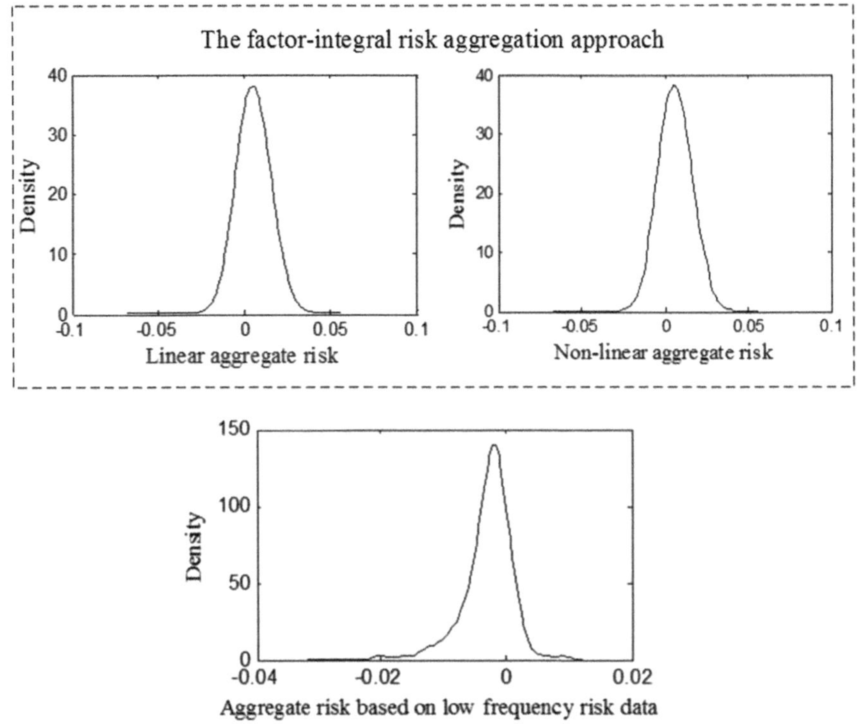

Fig. 5.6 The comparison of integrated loss distributions through the factor-integral approach and the low-frequency-risk data approach

From Fig. 5.6, it is clear that integrated loss distributions obtained through the linear and non-linear factor-integral approaches are similar, while the integrated loss distribution obtained using the low-frequency risk data approach is not smooth enough and is quite different from them. To be specific, the mean values of the integrated results obtained from the linear (0.0054) and non-linear (0.0062) factor integral models are positive. In contrast, the mean value of the integrated result obtained from the low-frequency-risk data approach is negative (−0.0030). The integrated loss from the non-linear factor-integral approach is moderately right-skewed at 0.0536, and that from the linear factor-integral approach is more right-skewed at 0.0759, while from the low-frequency-risk data approach, it is left-skewed at −1.5278. For volatility, the integrated loss from the linear factor-integral approach has the highest volatility (0.0103), while integrated loss based on low-frequency risk data has the lowest volatility (0.0042). Compared with the integrated loss from the low-frequency-risk data approach, which has the fattest tail (kurtosis = 5.8231), the relatively lower positive kurtosis values present in the linear (0.0886) and non-linear factor integral approaches (0.0611) suggest that integrated loss derived from the factor-integral approach has thinner tails.

The reason why the shape of integrated loss distribution from the low-frequency-risk data approach is significantly different from the integrated loss distributions derived by the factor-integral model is that the 807 pieces of quarterly risk P&L data are too small to fit the distribution very well. Adding or omitting several data might have a severe influence on the distribution of integrated loss. Therefore, the fitted integrated loss distribution from the low-frequency-risk data approach might differ from the actual distribution of integrated loss. In contrast, the integral factor approach uses high-frequency (as high as daily data) common risk factors such as equity market index and interest rate to model integrated loss, so the distribution of integrated loss depends on the distributions of risk factors. Since the high-frequency risk factor data are considered more accurate and stable when fitting a distribution, we can obtain more accurate distributions for common factors. Thus, the integrated loss distribution characterized by the risk factor distributions is more accurate and robust.

At the 99.9% confidence level, the 2019 annual integrated VaR and EC within the linear and non-linear factor-integral approaches and the low-frequency-risk data approach are recorded in Table 5.10 for comparison.

It is clear to see that the integrated VaR and EC obtained from different models are different. Since VaR is defined as a quantile of the distribution of risk returns, the larger negative value or smaller positive value of VaR corresponds to the higher level of risk (Rosenberg and Schuermann 2006). As shown in Fig. 5.6, integrated loss distribution based on low-frequency risk data has the fattest tail, so the integrated loss obtained from the low-frequency-risk data approach is the highest with the most significant negative value of VaR (−5.46%). The negative values of annual

Table 5.10 The comparison of 2019 annual integrated risk results from the factor-integral approach and the low-frequency-risk data approach

Integrated risk results (99.9% confidence level)	Proposed factor-integral approach		Low-frequency-risk data approach
	Linear	Non-linear	
VaR (%)	−5.12	−4.93	−5.46
EC (billion CNY)	5411	5210	4168

integrated VaRs obtained through the linear (−5.12%) and non-linear (−4.93%) factor-integral approaches are slightly smaller, so the integrated losses obtained via linear and non-linear factor-integral models are relatively lower.

However, EC is determined by integrated loss and expected loss. In other words, EC is not only decided by the tail but also dependent on the shape of the integrated loss distribution. The integrated VaRs recorded in Table 5.10, which determine integrated losses, are negative and closed to each other for both the factor-integral approach and the low-frequency-risk data approach. However, the mean values of integrated loss, which determine the expected losses are quite different from factor-integral approach to low-frequency-risk data approach. Specifically, the mean value of integrated loss from the low-frequency-risk data approach is negative (−0.00030), while the mean values of integrated loss from the linear (0.0054) and non-linear (0.0062) factor-integral approaches are positive (see Table 5.9). Therefore, EC values calculated through the factor-integral approach are different from the low-frequency-risk data approach.

Particularly, the values of EC obtained from factor-integral models, regardless of whether linear (5411 billion CNY) or non-linear (5210 billion CNY), are approximately 20% larger than the EC calculated through the low-frequency-risk data approach (4168 billion CNY) in 2019. In other words, on the basis of the factor-integral approach, the Chinese banking system should keep more EC to absorb potential unexpected losses arising from credit and market risks to ensure sound operations. The low-frequency risk data may lead to a biased shape of integrated loss distribution and thus an underestimated EC estimation.

Furthermore, we can also find from Table 5.10 that the difference between integrated results obtained from the linear and non-linear factor-integral approaches is slight. The reason may be that when applying our proposed factor-integral approach to the Chinese banking system, the linear risk factors can explain the majority of credit (R-squared = 76.78%) and market risk (R-squared = 63.74%). In non-linear regression, values of R-squared for credit risk and market risk increased to 87.55% and 76.81%, respectively. Thus the additional cross items only increase the explanatory power of risk factors for credit or market risk by approximately 10%. The slight growth in explanatory power may explain that the linear and non-linear factor-integral approaches resulted in similar integrated losses when used to aggregate Chinese banking credit and market risks.

5.5 Conclusions

In risk aggregation, most risk data are low-frequency, which leads to a problem of data shortage. The main contribution of this chapter is proposing a factor-integral approach to aggregating credit and market risks. The proposed approach can address the problem of risk data shortage to some extent by transforming the aggregation of low-frequency risk data into the integral of high-frequency risk factor data, which can obtain a more accurate and stable result of the integrated loss.

In the experiment, our proposed factor-integral approach is applied to the Chinese banking system to aggregate credit and market risks. To empirically analyze the difference between integrated results based on high-frequency risk factor data and low-frequency risk data, we further aggregate credit and market risks of the Chinese banking system based on low-frequency risk P&L data. Our empirical comparison proves that the integrated loss distribution based on low-frequency risk P&L data is quite different from our proposed factor-integral approach. Besides, compared with EC estimate based on low-frequency risk P&L data, the EC derived from our proposed factor-integral approach increases by approximately 25%, which indicates that banks need to keep more capital to protect against losses arising from credit and market risks.

Besides, our approach is applicable not only to aggregate credit and market risks but can also be easily extended to aggregate other risk types and more than two types of risks as long as these risks have common risk factors.

References

Aas K, Dimakos XK, Oksendal A (2007) Risk capital aggregation. Risk Manage 9(2):82–107. https://doi.org/10.1057/palgrave.rm.8250027

Adrian T (2017) Risk management and regulation. J Risk 20(1):23–57. https://doi.org/10.21314/JOR.2017.396

Alexander C, Pezier J (2003) On the aggregation of firm-wide market and credit risk. Henley Business School, Reading University

Basel Committee on Banking Supervision (2006) International convergence of capital measurement and capital standards: a revised framework. Bank for International Settlements

Basel Committee on Banking Supervision (2009) Findings on the interaction of market and credit risk. Bank for International Settlements

Basel Committee on Banking Supervision (2010) Developments in modelling risk aggregation. Bank for International Settlements

Bellini T (2013) Integrated bank risk modeling: a bottom-up statistical framework. Eur J Oper Res 230(2):385–398. https://doi.org/10.1016/j.ejor.2013.04.031

Bernal O, Gnabo JY, Guilmin G (2014) Assessing the contribution of banks, insurance and other financial services to systemic risk. J Bank Finance 47:270–287. https://doi.org/10.1016/j.jbankfin.2014.05.030

Bignozzi V, Tsanakas A (2014) Parameter uncertainty and residual estimation risk. J Risk Insurance 83(4):949–978. https://doi.org/10.1111/jori.12075

Bocker K, Hillebrand M (2009) Interaction of market and credit risk: an analysis of inter-risk correlation and risk aggregation. J Risk 11(4):3–29. https://doi.org/10.21314/JOR.2009.198

Breuer T, Jandacka M, Rheinberger K, Summer M (2010) Does adding up of economic capital for market and credit risk amount always to conservative risk estimates? J Bank Finance 34(4):703–712. https://doi.org/10.1016/j.jbankfin.2009.03.013

Brockmann M, Kalkbrener M (2010) On the aggregation of risk. J Risk 12(3):45–68. https://doi.org/10.21314/JOR.2010.208

Danielsson J, Zigrand JP (2006) On time-scaling of risk and the square-root-of-time rule. J Bank Finance 30(10):2701–2713. https://doi.org/10.1016/j.jbankfin.2005.10.002

Dionne G (2013) Risk management: history, definition, and critique. Risk Manage Insurance Rev 16(2):147–166. https://doi.org/10.1111/rmir.12016

Drehmann M, Sorensen S, Stringa M (2010) The integrated impact of credit and interest rate risk on banks: a dynamic framework and stress testing application. J Bank Finance 34(4):713–729. https://doi.org/10.1016/j.jbankfin.2009.06.009

Dunis CL, Laws J, Sermpinis G (2011) Higher order and recurrent neural architectures for trading the EUR/USD exchange rate. Quant Finance 11(4):615–629. https://doi.org/10.1080/14697680903386348

Furman E, Wang R, Zitikis R (2017) Gini-type measures of risk and variability: Gini shortfall, capital allocations, and heavy-tailed risks. J Bank Finance 83:70–84. https://doi.org/10.1016/j.jbankfin.2017.06.013

Grundke P (2009) Importance sampling for integrated market and credit portfolio models. Eur J Oper Res 194(1):206–226. https://doi.org/10.1016/j.ejor.2007.12.028

Grundke P (2010) Top-down approaches for integrated risk management: how accurate are they? Eur J Oper Res 203(3):662–672. https://doi.org/10.1016/j.ejor.2009.09.015

Grundke P (2013) On the reliability of integrated risk measurement in practice. J Risk 15(3):87–110. https://doi.org/10.21314/JOR.2013.260

Hartmann P (2010) Interaction of market and credit risk. J Bank Finance 34(4):697–702. https://doi.org/10.1016/j.jbankfin.2009.10.013

Imbierowicz B, Rauch C (2014) The relationship between liquidity risk and credit risk in banks. J Bank Finance 40:242–256. https://doi.org/10.1016/j.jbankfin.2013.11.030

Inanoglu H, Jacobs M (2009) Models for risk aggregation and sensitivity analysis: an application to bank economic capital. J Risk Finan Manage 2(1):118–189. https://doi.org/10.3390/jrfm2010118

Jarrow RA, Turnbull SM (2000) The intersection of market and credit risk. J Bank Finance 24(1):271–299. https://doi.org/10.1016/S0378-4266(99)00060-6

Kabaila P, Mainzer R (2018) Estimation risk for value-at-risk and expected shortfall. J Risk 20(3):29–47. https://doi.org/10.21314/JOR.2018.376

Kretzschmar G, McNeil AJ, Kirchner A (2010) Integrated models of capital adequacy-why banks are undercapitalized. J Bank Finance 34(12):2838–2850. https://doi.org/10.1016/j.jbankfin.2010.02.028

Kuritzkes A, Schuermann T (2007) What we know, don't know and can't know about bank risk: a view from the trenches. The known the unknown and the unknowable in financial risk management. Princeton University Press, Princeton

Li J, Feng J, Sun X, Li M (2012) Risk aggregation mechanisms and approaches in banking industry. Int J Inf Technol Decis Mak 11(6):1183–1213. https://doi.org/10.1142/S0219622012500320

Li J, Wei L, Lee CF, Zhu X, Wu D (2018) Financial statements based bank risk aggregation. Rev Quant Financ Acc 50(3):673–694. https://doi.org/10.1007/s11156-017-0642-0

Li J, Zhu X, Wu D, Lee CF, Feng J, Shi Y (2015) On the aggregation of credit, market and operational risks. Rev Quant Financ Acc 44(1):161–189. https://doi.org/10.1007/s11156-013-0426-0

Medova EA, Smith RG (2005) A framework to measure integrated risk. Quant Financ 5(1):105–121. https://doi.org/10.1080/14697680500117583

Nye CD, Drasgow F (2011) Assessing goodness of fit: simple rules of thumb simply do not work. Organ Res Methods 14(3):548–570. https://doi.org/10.1177/1094428110368562

Pianosi F, Wagener T (2015) A simple and efficient method for global sensitivity analysis based on cumulative distribution functions. Environ Model Softw 67:1–11. https://doi.org/10.1016/j.envsoft.2015.01.004

Rosenberg JV, Schuermann TA (2006) General approach to integrated risk management with skewed, fat-tailed risks. J Financ Econ 79(3):569–614. https://doi.org/10.1016/j.jfineco.2005.03.001

Wei L, Li G, Zhu X, Li J (2019) Discovering bank risk factors from financial statements based on a new semi-supervised text mining algorithm. Account Financ. https://doi.org/10.1111/acfi.12453

Wei Y, Wang Y, Huang D (2010) Forecasting crude oil market volatility: further evidence using GARCH-class models. Energy Econ 32(6):1477–1484. https://doi.org/10.1016/j.eneco.2010.07.009

Zhu X, Wand Y, Li J (2019) Operational risk measurement: a loss distribution approach with segmented dependence. J Oper Risk 14(1):1–20. https://doi.org/10.21314/JOP.2019.220

Zhu X, Wei L, Wu D, Li J (2018) A general framework for constructing bank risk data sets. J Risk 21(1):37–59. https://doi.org/10.21314/JOR.2018.393

Chapter 6
A Two-Stage General Approach Based on Financial Statements Data and External Loss Data

6.1 Introduction

Bank risk aggregation is still a great challenge because there are many types of bank risks and the basic characteristics of them vary greatly (Rosenberg and Schuermann 2006; Begley et al. 2016). Most existing studies only focus on aggregating two types of risks, and the studies that can aggregate three or more risks are scarce (Li et al. 2015). Effectively aggregating multiple types of bank risk is always a big challenge for two main reasons (Bernard and Vanduffel 2015; Begley et al. 2016). One is that there are many types of bank risks and the essential characteristics between multiple risks vary greatly (Rosenberg and Schuermann 2006). The other is that the complex correlations between different bank risks are hard to capture comprehensively (Hartmann 2010). The multiple correlation structures between different types of risks show various complex characteristics, such as nonlinearity, tail dependence and tail asymmetry, making it difficult to accurately model the bank risk correlations (Li et al. 2015).

There are two major kinds of risk aggregation approaches so far (Hartmann 2010; Baker and Filbeck 2015). One is the top-down approach (TDA), where the marginal distributions of individual risks are derived separately and then linked by correlation matrices or copula functions to model their joint distribution function (Basel Committee on Banking Supervision, BCBS for short, 2010; Li et al. 2015). TDAs have a broader scope of application because they are intuitive and easy to use. However, whether these matrices or copulas can sufficiently represent the complex interactions between various risk types might be questionable (Grundke 2010; Grundke and Polle 2012). An alternative way to aggregate risks is to follow the bottom-up approach (BUA), which models the complex interactions between different risk types on the level of risk factors (Bocker and Hillebrand 2009). BUAs are always considered to be more accurate than TDAs because they determine the correlation relationship between risk types naturally from their origins (Grundke

J. Li et al., *Financial Statements-Based Bank Risk Aggregation*, Innovation in Risk Analysis, https://doi.org/10.1007/978-981-19-0408-0_6

2010; Bellini 2013). However, BUAs are designed to aggregate bank risks with common risk factors, which means risks without common risk factors cannot be aggregated using BUAs.

As discussed above, using the top-down approach to aggregate all these risks is feasible. Nevertheless, when the number of risks becomes large, the complexity of the widely-used copulas such as Gaussian copula and t copula will significantly increase so that their capability of capturing dependence structure will deteriorate (Okhrin and Tetereva 2017; Zhu et al. 2020). Therefore, the accuracy of aggregate results is even more doubtable. BUAs are more accurate and reasonable, but some bank risks, such as operational risk, are unique. It is far from clear which observable risk factors might drive them (Rosenberg and Schuermann 2006). Thus, at present, there is still a lack of a bank risk aggregation approach that can accurately aggregate multiple bank risks.

Credit, market, and operational risks are widely recognized as the three major risks banks face (BCBS 2006; Wei et al. 2019). Among these three types of risks, the credit risk and market are generally considered to have many common factors, while the operational risk does not have obvious factors (Bocker and Hillebrand 2009). Thus, by establishing the relationship between bank risk profit and loss (P&L) and risk factors, we can measure credit risk and market risk.

Income statements record bank profit and loss (P&L) in their daily business. Since earnings and losses arising from bank risks within specific periods are recorded and summarized in the income statement, looking for risk proxies from financial statements provides a reasonable basis for decomposing earnings and losses into risk sources. There have been a number of studies, for instance, Rosenberg and Schuermann (2006), Kuritzkes and Schuermann (2007) and Inanoglu and Jacobs (2009), suggesting that mapping income statement items into risk types is an alternative way to collect risk data. The specific mapping relationships in these studies are shown in Table 6.1.

As shown in Table 6.1, the mapping between market risk and trading income is the clearest. Market risk refers to the potential losses arising from changes in the value or price of an asset, such as those resulting from fluctuations in interest rates, currency exchange rates, stock prices and commodity prices (BCBS 2006). Fittingly,

Table 6.1 The mapping between income statement items and bank risks in previous studies

Studies	Mapping relationship		
	Credit risk	Market risk	Operational risk
Rosenberg and Schuermann (2006)	Net interest income – provisions	Trading revenue	–
Kuritzkes and Schuermann (2007)	– provisions	Trading income	40% × (Other income + Non-interest expense + net extraordinary items)
Inanoglu and Jacobs (2009)	Gross charge-offs	Net trading revenues	Non-interest expense

the item of trading income records the earnings resulting from the bank's market risk-related activities. With respect to credit risk, taking provisions as the risk proxy is a good choice. Credit risk is defined as the potential loss caused by borrower's or counterparty's failure to meet its obligations. Hence, the item of provisions, which is a charge for incurred loan losses arising from credit default, is inherently related to the definition of credit risk. Moreover, a robustness check presents that the choice between provision and charge-offs appears to make little difference to measure credit risk (Kuritzkes and Schuermann 2007).

However, the mapping relationship between operational risk and income statement items is rough. Operational risk is defined as the risk of direct or indirect loss resulting from inadequate or failed internal processes, people and systems, or external events (BCBS 2006). Thus, some researchers simply mapped the remaining income statement items that are not mapped into credit and market risks into operational risk. Based on the definition of operational risk, some researchers found that losses from operational risk events are mostly recorded under the item of non-interest expenses. Besides, Inanoglu and Jacobs (2009) also mapped operational risk into non-interest expenses.

Actually, the mainstream is to measure operational risk through the external database. External databases contain operational loss events. Many studies measure operational risk events based on external loss events. So in the process of aggregating risk, there are other data sources besides financial statements. We can use external loss data to supplement financial statements data.

Based on the above discussions, however, the existing two types of risk aggregation approaches have limitations, and thus prevent them from being an excellent way to solve the difficulties faced by bank risk aggregation. Specifically, TDAs can be easily used to aggregate multiple bank risks, but they are incapable of sufficiently capturing the complex interactions between risk types and may lead to biased aggregate results. BUAs are more accurate than TDAs by capturing the dependence between risks based on the interaction of common risk factors. However, they are not applicable to aggregating multiple bank risks without common risk factors. These limitations highlight that there are few aggregation approaches that can effectively aggregate multiple bank risks. This is the issue that we address in this chapter.

Therefore, we propose a new two-stage general risk aggregation framework to effectively aggregate credit, market and operational risk based on financial statements and the external database. Specifically, by reasonably combing the TDAs and BUAs in a framework, multiple bank risks with different characteristics can be aggregated in an accurate way. The basic idea of the approach is to first divide risks according to whether they have common factors. We give priority to risks with common risk factors by aggregating them through a BUA in the first stage based on financial statements. Then, the subtotal risk obtained by the BUA and the remaining risks without common risk factors is further aggregated using a TDA in the second stage based on the external database.

The rest of this chapter is structured as follows. Section 6.2 illustrates the proposed two-stage bank risk aggregation approach and its specific implementation steps. Section 6.3 describes the data used in this chapter. In Sect. 6.4, the proposed approach is applied to aggregating the credit, market and operational risks of the Chinese banking sector. Section 6.5 summarizes the conclusions.

6.2 A Two-Stage Risk Aggregation Approach

The two-stage general risk aggregation approach proposed by us is used to aggregate credit, market and operational risks with different databases. Figure 6.1 provides a complete outline of our proposed framework. Specifically, among the n types of risks that we want to aggregate. First, we classify them into two categories based on whether they have common risk factors. Then, these two categories of risks are aggregated through two stages. Specifically, risks with common risk factors $(Risk_1, Risk_2, \ldots, Risk_m)$ are aggregated using the BUA in the first stage. The reason is that the BUA is deemed to be more accurate than the TDA. Therefore, the BUA is the choice for risks with common risk factors. In the second stage, the aggregate risk obtained by the BUA and the residual risks $(Risk_{m+1}, Risk_{m+2}, \ldots, Risk_n)$ without common risk factors are aggregated using the TDA to finally obtain the total risk through $n - m$ aggregation levels.

Therefore, our proposed two-stage general risk aggregation framework can aggregate various risks through two stages with a BUA and a TDA. The BUA and TDA embedded in this framework are the factor-integral approach and copula-based hierarchical aggregation approach, respectively. We then introduce each of them in detail.

6.2.1 *Stage 1: The Factor-Integral Approach for Financial Statements Data*

The BUA embedded in our proposed framework is the factor-integral approach, which was initially proposed for aggregating credit and market risks (Wei et al. 2018). In fact, the factor-integral approach can be extended to aggregate bank risks as long as they have common risk factors. Hence, here we extend the factor-integral approach and embed it into our proposed framework to aggregate bank risks with common risk factors. In the following, we provide a detailed introduction to clarify the aggregation steps of using the extended factor-integral approach to aggregate risks with common risk factors.

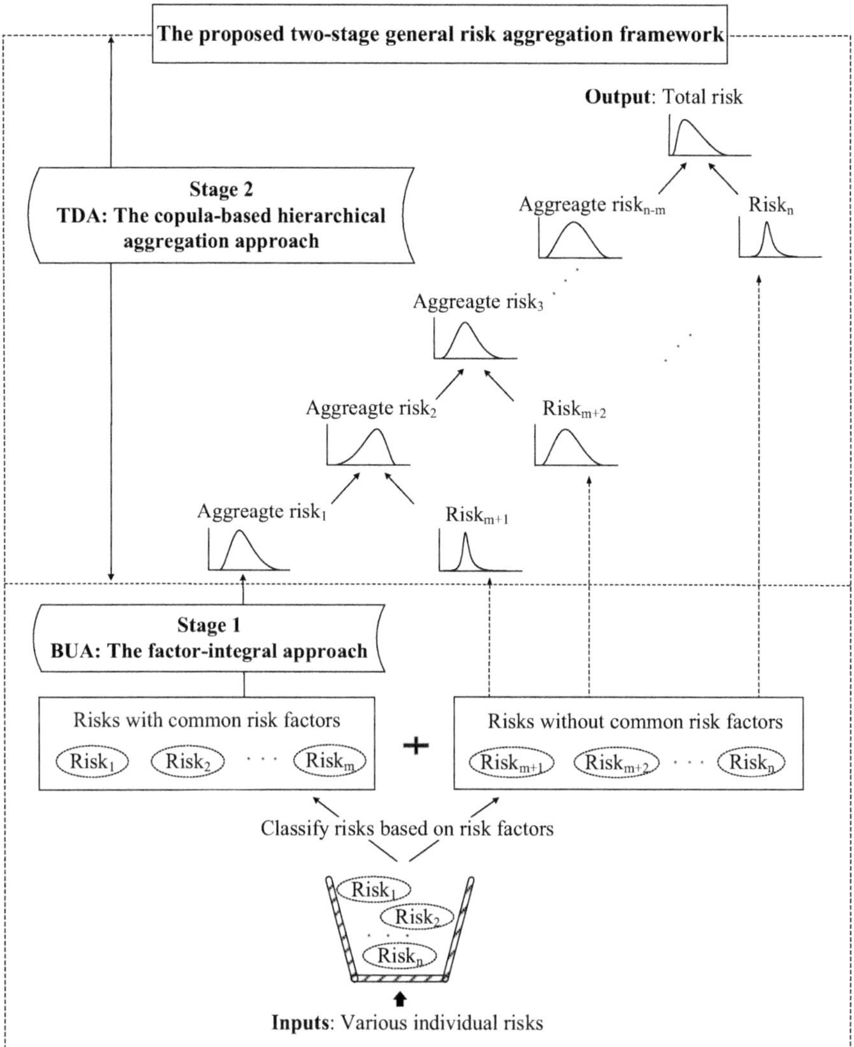

Fig. 6.1 The whole process of using our proposed two-stage general risk aggregation framework

6.2.1.1 The Concept of the Factor-Integral Approach

The factor-integral approach aggregates bank risks from the view of their underlying common risk factors. Thereby it belongs to BUAs. Specifically, movements in common risk factors will affect different types of risks and finally lead to an aggregate P&L. Different risks are naturally correlated based on their common underlying factors (Jarrow and Turnbull 2000; Grundke 2010). Thus, dependencies between

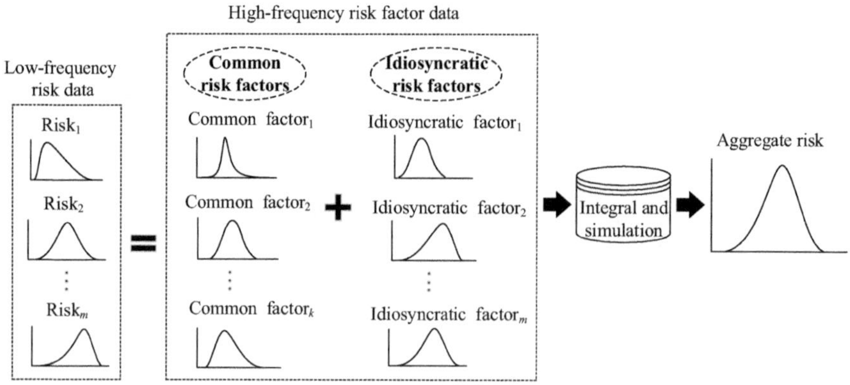

Fig. 6.2 The illustration of the BUA called the factor-integral approach

different risk types are considered on the risk factor level in the aggregate risk estimate because the aggregate risk is determined by movements in common risk factors (Grundke 2010). Therefore, the factor-integral approach is a BUA and crucially depends on two issues: the identification of common risk factors and the appropriate fitting distributions of common risk factors (Bocker and Hillebrand 2009; Breuer et al. 2010).

The extension concept of the factor-integral approach is intuitively illustrated in Fig. 6.2. To be specific, for risks that have common risk factors, their individual risk profit and loss (P&L) is substantially explained by common risk factors, and the residual is explained by its idiosyncratic risk factor as uncertainties about risk P&L are related through sensitivities to uncertainties about risk factors (Alexander and Pezier 2003; Rosenberg and Schuermann 2006). Having identified common risk factors, we can directly construct the cumulative probability density distribution of aggregate risk through multiple integrals of high-frequency risk factors based on Monte Carlo simulation.

In general, the factor-integral approach for m risks $Risk_1, Risk_2, \ldots, Risk_m$ shown in Fig. 6.1 that are affected by k common risk factors $(X_1, X_2, \ldots X_k)$ can be modeled as follows.

Let L^i denote the P&L of individual risk i. Let $\beta_1^i, \beta_2^i, \ldots, \beta_k^i$ denote the parameters of common risk factors. The formulation between risk P&L and risk factors is written as:

$$L^i = F_i(X_1, X_2, \ldots, X_k) + \varepsilon^i \tag{6.1}$$

A simple solution is a linear regression model. Hence, we rewrite L^i from (Eq. 6.1) as:

$$\begin{aligned} L^i &= \alpha^i + \beta^i \mathbf{X} + \varepsilon^i \\ &= \alpha^i + \beta_1^i X_1 + \beta_2^i X_2 + \ldots \beta_k^i X_k + \varepsilon^i \end{aligned} \tag{6.2}$$

It may be inappropriate to assume a linear relationship between risk P&L and risk factors. In most forms, risk P&L is nonlinearly related to risk factors. However, an extreme variation that only occurred in a factor with minimal sensitivity would not induce an extreme P&L variation (Alexander and Peizer 2003). Furthermore, higher order models have higher computational costs and may cause an over-fitting problem (Dunis et al. 2011). Thus, in a much more general setting, a second-order Tailor expansion approximation is given by:

$$L^i = \alpha^i + \beta^{\mathbf{i}}\mathbf{X} + 1/2\mathbf{X}'\gamma^{\mathbf{i}}\mathbf{X} + \varepsilon^i = \alpha^i + \sum_{a=1}^{k}\beta_a^i X_a + 1/2\left[\sum_{a=1}^{k}\gamma_{aa}^{i2}X_a^2 + 2\sum_{a,b}\gamma_{a,b}^i X_a, X_b\right] + \varepsilon^i \quad (6.3)$$

Having fixed the relationship between individual risk P&L and risk factors, we then calculate the aggregate risk values and their corresponding probabilities. Let L represent the aggregate risk (Eqs. 6.4 and 6.8) can be derived based on (Eq. 6.1).

$$P(L \le r) \quad (6.4)$$

$$= P\left(L^1 + L^2 + \cdots + L^m \le r\right) \quad (6.5)$$

$$= P\left[F_1(X_1, X_2, \ldots, X_k) + \cdots + F_m(X_1, X_2, \ldots, X_k) + \varepsilon^1 + \cdots + \varepsilon^m \le r'\right] \quad (6.6)$$

$$= \iint \ldots \int_D f\left(\varepsilon^1\right)d\varepsilon^1 \ldots f\left(\varepsilon^m\right)d\varepsilon^m f(X_1)dX_1 \ldots f(X_k)dX_k. \quad (6.7)$$

where the integral region D is that $X_1, X_2, \ldots, X_k$ satisfied a condition like

$$F_1(X_1, X_2, \ldots, X_k) + \cdots + F_m(X_1, X_2, \ldots, X_k) + \varepsilon^1 + \cdots + \varepsilon^m \le r \quad (6.8)$$

When common risk factors are continuous variables, from (Eqs. 6.4 and 6.8), given specific aggregate risk value, its corresponding probability can be calculated through multiple integrals of risk factors. A large number of aggregate risk values and their corresponding probabilities allow us to construct the empirical cumulative probability density distribution of the aggregate risk. Thus, this BUA is called the factor-integral approach. In practice, the common risk factors are discrete variables. Based on plenty of Monte Carlo simulations of common risk factors, we can simulate realizations of aggregate risk $L = L^1 + L^2 + \ldots + L^m$ with $L^1, L^2, \ldots, L^m$ defined as in Eqs. (6.2) and (6.3). By doing this, an empirical cumulative density function for the aggregate risk L can be generated.

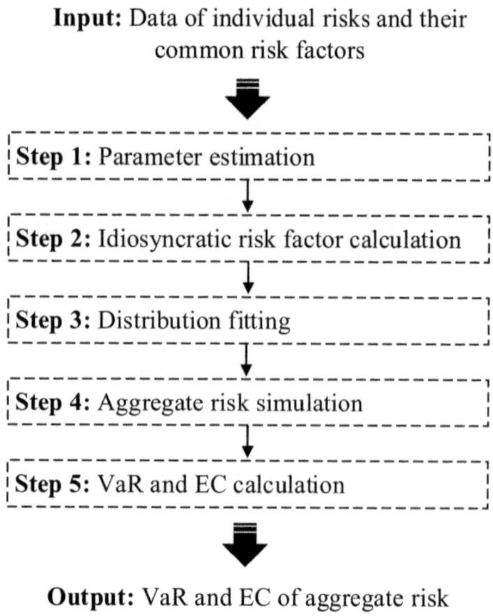

Fig. 6.3 The specific steps of using the factor-integral approach

6.2.1.2 Specific Steps of the Factor-Integral Approach

Figure 6.3 gives the specific steps of using the extended factor-integral approach. Determine the data of individual risk P&L and their common risk factors as input. It is noteworthy that the common risk factor data must be adjusted to the same frequency as the risk P&L data. The initial input can be transformed into the final output of aggregate value-at-risk (VaR) and EC following the five steps.

Input: Data of $L^1, L^2, \ldots, L^m$ and $X_1, X_2, \ldots, X_k$.

Step 1 Parameter estimation

First, we need to estimate the parameters of the function $L^i = F_i(X_1, X_2, \ldots, X_k) + \varepsilon^i$.

Step 2 Idiosyncratic factor calculation

The idiosyncratic factor data ε^i can be calculated by using (Eq. 6.1).

$$\varepsilon^i = L^i - F_i(X_1, X_2, \ldots, X_k) \tag{6.9}$$

Step 3 Distribution fitting

In this step, the best-fitting distributions $f(X_1), f(X_2), \ldots, f(X_k)$ for common risk factors and $f(\varepsilon^1), f(\varepsilon^2), \ldots, f(\varepsilon^m)$ for idiosyncratic factors are found by the widely used goodness-of-fit Kolmogorov–Smirnov (KS) test (Bernal et al. 2014). In addition, estimations of the factor distributions' parameters are determined through the maximum likelihood method. As a robustness check, an alternative goodness-of-fit chi-square test is also adopted (Bryant and Satorra 2012).

Step 4 Aggregate risk simulation

A large number of simulations are needed to construct the distribution of aggregate risk. In each simulation, given an r value, its corresponding probability is calculated by (Eq. 6.7). A larger number of simulations will result in the simulated aggregate risk distribution being closer to the actual distribution and a longer required computational time. To balance simulation accuracy and costs, the simulation times N is widely considered to meet the requirement of (Eq. 6.10).

$$N \times \alpha \times (1 - \alpha) \geq 50 \tag{6.10}$$

where α represents the confidence level. This chapter is prepared to calculate VaR at the confidence levels of 95, 98, 99 and 99.9%. Thus, we set N as 100,000 here, which conforms to (Eq. 6.10).

Step 5 VaR and EC calculation

After obtaining the aggregate risk distribution, we employ the Value-at-Risk (VaR), which has become a standard measure for aggregate risk (Chen and Tu 2013; Peng et al. 2019). VaR at a specific confidence level $a \in (0, 1)$ is defined as the smallest number l such that the probability of loss L exceeding l is not larger than $1 - \alpha$:

$$\text{VaR} = \inf\{l : P(L \leq l) \leq (1 - \alpha)\} \tag{6.11}$$

To have an idea of how much capital should be available to absorb the potential unexpected losses from dependent risks, we further calculate EC, which is defined as the difference between total losses and expected losses for the aggregate risk distribution (Alessandri and Drehmann 2010; Furman et al. 2017).

Output: VaR and EC of aggregate risk distribution

The above five specific steps clearly describe how to use the factor-integral approach to aggregate risks with common risk factors. Within our proposed framework, it is feasible and applicable to employ the factor-integral approach to aggregate risks with common risk factors in the first aggregation stage.

6.2.2 Stage 2: The Copula-Based Hierarchical Aggregation Approach for External Data

To aggregate risks without common risk factors in the second aggregation stage, the TDA embedded in our proposed framework is the copula-based hierarchical aggregation approach. Although the copula-based hierarchical aggregation approach has been used in some studies (Yang et al. 2015; Okhrin and Tetereva 2017), to our best knowledge, it has not been applied to aggregating bank risks. Thus, we first introduce the copula-based hierarchical aggregation approach into the field of bank risk aggregation.

As discussed in the introduction section, copula-based approaches are the most promising trend for TDAs (Bocker and Hillebrand 2009). For aggregating multiple risks, the multivariate copula requires to specify the copula of all risks. However, it is often problematic when used in high dimensions because the dependence structures are too complex to accurately model. Thus, fitting the multivariate copula models is problematic, and copula model simulation is numerically slow in high dimensions (Arbenz et al. 2012). The copula-based hierarchical aggregation approach can overcome the limitations of multivariate copula models and is even applicable in high dimensions. It aggregates the risks hierarchically, allowing the use of different bivariate copula to model dependence characteristics for each aggregation step. In practice, the model complexity and the number of parameters can be adjusted to the situation at hand. To have a basic knowledge of the copula-based hierarchical aggregation approach, researchers can refer to Arbenz et al. (2012).

In general, the copula-based hierarchical aggregation for p risks consists of a tree structure, p-1 bivariate copulas, and p marginal distributions (Côté and Genest 2015). Here, we assume that three risks T, X, Y (where T is an aggregate risk that is obtained through the factor-integral approach introduced in Sect. 6.2.1) serve as inputs to illustrate how to use the copula-based hierarchical aggregation approach. As shown in Fig. 6.4, we can finally derive the total risk VaR and EC based on the following five steps.

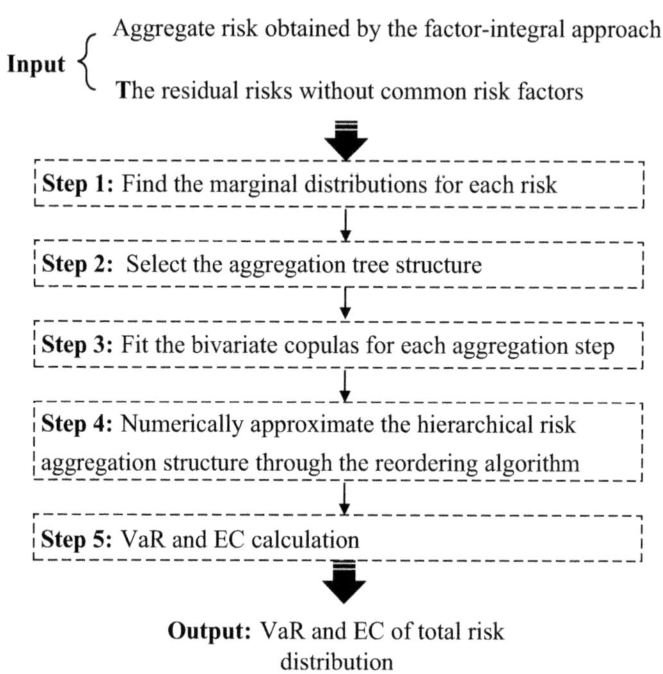

Fig. 6.4 The procedure of the copula-based hierarchical aggregation approach

Step 1 Find the marginal distributions for each risk.

Marginal distributions for each of these three risks T, X, Y are estimated and given by:

$$F_T(t) = P[T \le t] \ F_X(x) = P[X \le x] \ \ F_Y(y) = P[Y \le y] \tag{6.12}$$

Step 2 Select the aggregation tree structure.

The aggregation tree structure is determined based on risk dependence. The most correlated risks are aggregated at the same tree level. Specifically, Risk X that is most dependent on the risk T based on Kendall's tau τ, is selected to be aggregated with risk T in the first aggregation level. Next, the aggregate risk Z of risk T and X is aggregated with the remaining risk Y at the second aggregation level. Thus, the aggregation tree for these three T, X, Y risks is accordingly determined.

Step 3 Fit the bivariate copulas for each aggregation step.

For the candidate copula families, goodness-of-fit tests proposed by Kojadinovic and Yan (2010) are employed to select the best-fitted copula. In the first aggregation level, we model the dependence structure between T and X with a bivariate copula $C_{T,X}$.

$$P[T \le t, X \le x] = C_{T,X}(F_T(t), F_X(x)) \tag{6.13}$$

Therefore, the cumulative distribution function (F_Z) of aggregate Z is determined and given by:

$$F_Z(z) = P[T + X \le z] = \int_{R^2} C\{t + x \le z\} dC_{T,X}(F_T(t), F_X(x)) \tag{6.14}$$

where $\{t + x \le Z\}$ is an indicator function defined as:

$$C\{t + x \le z\} = \begin{cases} 1 \ if \ t + x \le z \\ 0 \ if \ t + x > z \end{cases} \tag{6.15}$$

After getting the distribution of aggregate risk Z in the previous step, we then combine F_Z and F_Y using the bivariate copula $C_{Z,Y}$ to determine the distribution of the total risk S.

$$P[Z \le Z, Y \le y] = C_{Z,Y}(F_Z(z), F_Y(y)) \tag{6.16}$$

$$S = Z + Y = (T + X) + Y \tag{6.17}$$

The cumulative distribution function (F_S) of total risk S is written as:

$$F_S(s) = P[Z + Y \le s] = \int_{R^2} C\{z + y \le s\} dC_{Z,Y}(F_Z(z), F_Y(y)) \tag{6.18}$$

Step 4 Numerically approximate the hierarchical risk aggregation structure through reordering algorithm.

Simulate marginal samples of size q for T, X, Y, $C_{T,X}$, $C_{Z,Y}$, respectively.

$$T_i \sim F_T \; X_i \sim F_X \; Y_i \sim F_Y \; U_i \sim C_{T,X} \; V_i \sim C_{Z,Y} \quad \text{for i} = 1, \ldots, \text{q}$$

Then, in the first aggregation level, reorder the marginal samples T_i and X_i based on the copula sample U_i to construct a bivariate reordered samples of (T, X). By doing this, we get a sample of Z by $Z = T + X$.

In the same way, we construct the bivariate reordered samples of (Z, Y) by reordering the marginal samples Z_i and Y_i based on the copula sample V_i. The sample of total risk S by $S = Z + Y$ in the second aggregation level determines the empirical distribution of the total risk S.

Step 5 VaR and EC calculation.

In our proposed framework, we adopt VaR and EC to measure the total risk based on the total risk distribution derived in Step 4.

Overall, starting from classifying risks into two categories based on whether they have common risk factors, we propose a two-stage general risk aggregation framework from a new risk aggregation perspective. Theoretically, having determined various risks as inputs, a more accurate total risk will be derived using a BUA for the first aggregation stage and a TDA for the second aggregation stage. Thus, the proposed framework has the advantages of both generality and accuracy.

6.3 Data Description

In the use of our proposed framework, the data on the individual risks and common risk factors should be assembled. Specifically, credit and market risk P&L data are obtained from financial statements by mapping risk types into income statement items. According to previous studies, the proxies of credit and market risks are summarized in Table 6.2 (Rosenberg and Schuermann 2006; Kuritzkes and Schuermann 2007; Li et al. 2017).

As discussed in Chap. 4, we collected quarterly credit and market risk data from 16 early listed A-share Chinese commercial banks (Table 4.1). After eliminating missing data, we finally obtained 807 observations of valid quarterly data for credit

Table 6.2 The proxies of credit risk and market risk from financial statement

Risk type	Proxy
Credit risk	– Provision/RWA
Market risk	(Net investment income + net foreign exchange differences + the changes in fair value gains or losses)/RWA

and market risks from 2007 to 2019. In 2019, the total risk-weighted assets (RWA) of 16 listed Chinese commercial banks were 105,694 billion CNY.

For high-frequency data of common risk factors for credit and market risks, we describe the data collection process in previous Sect. 5.3.2. Specifically, previous studies have determined that the interest rate, equity index, credit spread, exchange rate, commodity index, macroeconomic variables and their volatilities commonly affect credit and market risks (Alexander and Peizer 2003; Grundke 2010). Table 5.3 in Sect. 5.3.2 records the descriptions of the specific indicators of these six types of common risk factors. All data on common risk factors are from the Wind database (http://www.wind.com.cn/) and span from 2007 to 2019.

To capture the quarterly volatility features of common factors, The generalized autoregressive conditional heteroskedasticity (GARCH) (1,1) approach, a widely used and classical method that has worked well in most applied situations is employed by us (Wei et al. 2010). To meet the stationarity requirement for the GARCH analysis, the measures of common risk factors are the Ln-differences from the original data series. In the regression, to match the frequency of the credit and market risk returns derived from quarterly financial statements, daily factors have to be transformed into quarterly data by summing data within a quarter (Rosenberg and Schuermann 2006).

Unlike credit and market risks, operational loss data usually come from external operational loss databases (Biell and Muller 2013). The prominent operational loss databases include ORX, SAS OpRisk Global Data and others (Brechmann et al. 2014). To analyze operational risk in the Chinese banking system, we collect operational loss data from the Chinese Operational Loss Database (COLD), which is constructed by our research team. It is the largest operational risk database in China and has already been successfully applied to several published studies (Li et al. 2014a, b). However, the Chinese Operational Loss Database (COLD) just contains 2132 operational loss events spanning from 1994 to 2012. Due to data limitation, the operational risk loss here is measured based on historical operational risk events that occurred between 1994 and 2012.

6.4 Empirical Results

6.4.1 Aggregating Results of Credit and Market Risk

Among credit, market and operational risks, economic theory tells us that market and credit risks are commonly affected by some risk factors (Jarrow and Turnbull 2000). Based on our proposed framework, we obtain the subtotal risk of credit and market risks through the factor-integral approach in the first aggregation stage. Here, we adopt the most general form of the factor-integral approach. We establish a non-linear relationship between risk P&L and risk factors through the second-order Tailor expansion (Eq. 6.3).

Table 6.3 The subtotal risk of credit risk and market risk of Chinese banks

Subtotal risk	95%	98%	99%	99.9%
VaR (%)	−1.91	−2.76	−3.28	−4.93
EC (billion CNY)	2019	2917	3467	5210

Note The subtotal risk VaR directly calculated from the distribution is quarterly because the credit and market risk data is quarterly. To transform the quarterly VaR to the annual VaR, the quarterly VaR is multiplied by the square root of 4 (4 quarters in a year) based on a commonly used square-root-of-time rule scaling method from Danielsson and Zigrand (2006) and Aas et al. (2007)

By using the factor-integral approach, the subtotal risk distribution of the aggregated credit risk and market risk can be obtained. Then the Value-at-Risk (VaR), a standard measure used in financial risk management (Hull 2012; Li et al. 2020), is used to calculate the magnitude of subtotal risk based on its distribution.

The annual subtotal risk VaR and EC of the Chinese banking system are presented in Table 6.3. At the 99.9% confidence level that BCBS recommends, the subtotal risk VaR is −4.93%. In 2019, the RWA of the 16 listed Chinese banks was 105,694 billion CNY. So it is estimated that these Chinese banks should keep 5210 billion CNY in total to protect against the potential losses from credit and market risks in 2019.

6.4.2 *Operational Risk Measurement Results*

The operational risk is distinctly different from credit and market risks in terms of common risk factors. Credit risk and market risk are affected by some common risk factors, while operational risk shares no common risk factors with them. Thus, based on our proposed framework, operational risk is aggregated through the TDA in the second aggregation stage. Operational risk distribution should be obtained first for further aggregation. Therefore, in this section, we adopt the loss distribution approach (LDA), a popular methodology for measuring operational risk to construct the operational risk distribution of the Chinese banking system based on the COLD (Li et al. 2014a, b).

The LDA separately estimates the frequency distribution and severity distribution of operational risk loss and then derives the final distribution of operational risk loss through a Monte Carlo simulation. Concerning the loss frequency, Poisson, negative binomial and geometric distributions are frequently used (Chapelle et al. 2008). With respect to the loss severity, the lognormal distribution, the generalized hyperbolic distribution (GHD), the generalized error distribution (GED), the skewed generalized error distribution (SGED) and the generalized Pareto distribution (GPD) are widely used (Feng et al. 2012). Thus, we adopt the KS test here to determine which distribution type best fits the frequency and severity distributions.

The KS test results and parameters estimated by the maximum likelihood methods for frequency and severity distributions are presented in Tables 6.4 and 6.5, respectively. Table 6.5 illustrates that the negative binomial distribution outperforms the

Table 6.4 Estimated parameters and KS test results of the frequency distribution

Distribution	Parameters	D value	P value
Poisson	$\lambda = 112.21$	0.45	0.00
Negative binomial	$r = 2.5,\ p = 0.98$	0.20	0.43
Geometric	$d = 0.01$	0.28	0.10

Table 6.5 Estimated parameters and KS test results of the severity distribution

Distribution	Parameters	KS test	
		D value	P value
Lognormal	$u = 5.29, \sigma = 3.21$	0.04	0.00
GED	$u = 5.43, \sigma = 3.21, \upsilon = 3.51$	0.04	0.00
SGED	$u = 5.41, \sigma = 3.28, \upsilon = 3.68, \xi = 1.13$	0.03	0.07
GPD	$\xi = 3.35, \beta = 35.91$	0.08	0.00
GHD	$\alpha = 2.42, \beta = 0.55, \delta = 13.76, \mu = 0.00, \lambda = 16.45$	1.00	0.00

Table 6.6 The 2019 annual operational risk of the Chines banking system

Operational risk	95%	98%	99%	99.9%
VaR (%)	−0.05	−0.06	−0.07	−0.12
EC (billion in CNY)	53	64	74	127

other two frequency distributions as it has the most significant p-value, which is far greater than 5%. From Table 6.6, only the p-value of SGED is greater than 5%. Therefore, the SGED performs best on fitting severity distribution.

Having specified the marginal frequency distribution and marginal severity distribution, the annual loss distribution of the operational risk can be obtained through the Monte Carlo simulation based on the LDA. To remain consistent with the stated market and credit risks, the operational risk losses are adjusted (as with the risk returns) by dividing by the total RWA of the 16 Chinese commercial banks. Since the credit and market risks are measured based on P&L data, the value of the operational loss event and the operational risk return data are negative.

Table 6.6 records the VaR and EC of the operational risk at different confidence levels. The shape of the operational risk distribution is visually shown in Fig. 6.5. For example, we can see from Table 6.6 that at the 99.9% confidence level, the operational risk loss return of the Chinese banking system was −0.12%, and the corresponding EC was 127 billion CNY in 2019.

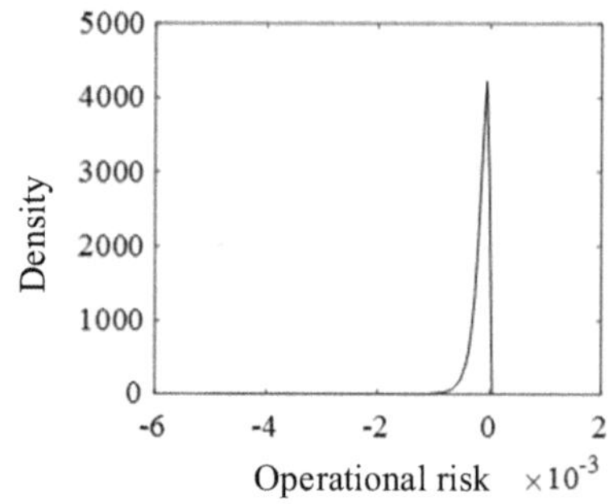

Fig. 6.5 The operational risk distribution of the Chinese banking sector

6.4.3 Total Risk Results

Based on our proposed framework, having determined the subtotal risk of credit and market risks with common risk factors and the residual operational risk without common risk factors, we can finally obtain the total risk using the TDA of copula-based hierarchical aggregation approach.

According to the specific steps of using the copula-based hierarchical aggregation approach, we first need to find the best-fitted distributions of annual subtotal risk and operational risk using the KS and chi-square tests. As recorded in Table 6.7, the subtotal risk follows a normal distribution, and the operational risk follows the Weibull distribution with three parameters.

The frequently used elliptical copulas and Archimedean copulas are considered as candidate copulas to model the dependence between subtotal risk and operational risk. The goodness-of-fit test for these copulas adopted here is proposed by Kojadinovic and Yan (2010). Just as the KS test, the larger the P-value is, the better the copula can fit the dependence structure. The results show that Gaussian copula with the parameter of 0.64 reaches the most significant P-value, so it is employed to aggregate the subtotal risk and operational risk. Having determined the bivariate copula for subtotal risk and operational risk, we can obtain total risk distribution through the reordering algorithm. The total risk distribution derived from our proposed framework is shown in Fig. 6.6. The total risk results at different confidence levels are shown in Table 6.8. At the 99.9% confidence level, the annual total risk VaR obtained from

Table 6.7 The fitted distribution of subtotal risk and operational risk

Risk type	Distributions	Parameters	KS test	
			D value	P-value
Sub-total risk	Normal	$u = 0.13, \sigma = 0.02$	0.24	0.39
Operational risk return	Weibull	$\lambda = 98.50, \kappa = 0.01, \gamma = 0.01$	0.17	0.26

Note For the normal distribution, u is the mean and σ is the standard deviation. For Weibull distribution, λ is scale parameter, K is shape parameter, and γ is position parameter

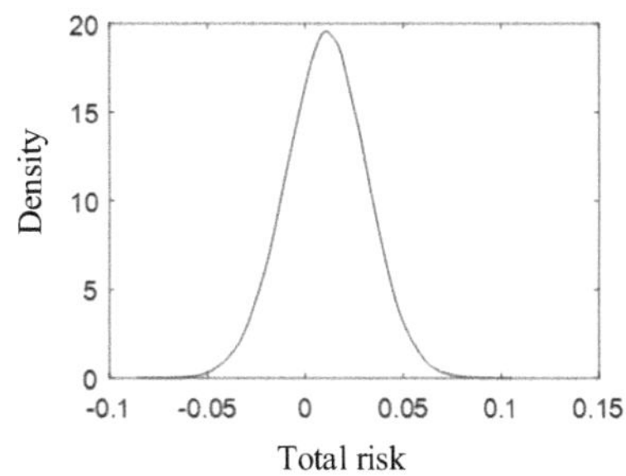

Fig. 6.6 The distribution of total risk derived from our proposed framework

Table 6.8 The total risk of the Chinese banks

Total risk	95%	98%	99%	99.9%
VaR (%)	−2.65	−3.40	−3.86	−5.37
EC (billion CNY)	2801	3594	4080	5676

our proposed approach is −5.37%, which means the Chinese banking sector should keep 5676 billion CNY to protect against the potential total losses caused by credit, market, and operational risks.

6.4.4 Results Comparisons

At last, the proposed approach is compared with three other widely used risk aggregation approaches in terms of total risk and diversification benefit. Since there is no BUA that can simultaneously aggregate credit, market, and operational risks, three top popular TDAs, i.e. simple summation approach, variance–covariance approach, and copula approach are used for comparison (BCBS 2010; Li et al. 2015).

Under TDAs, first, individual risks should be separately measured. Following Rosenberg and Schuermann (2006), the P&L distribution for credit and market risks can be derived based on their risk factors, respectively. The operational risk distribution has already been obtained in Sect. 6.4.2. The marginal risk distributions of the three risks are shown in Fig. 6.7. Then the VaRs of them at different confidence levels are calculated and shown in Table 6.9. At the 99.9% confidence level, the annual VaRs of these three risks are −3.99%, −5.08%, and −0.12%, respectively.

The total risk VaRs from our proposed approach and three comparison approaches are summarized in Table 6.9. The diversification benefit is measured by diversification coefficient, defined as the savings between the aggregated VaR and the sum of single VaRs for each risk (Brockmann and Kalkbrener 2010; Li et al. 2015). The diversification coefficients of these approaches at the 99.9% confidence level are graphically shown in Fig. 6.8.

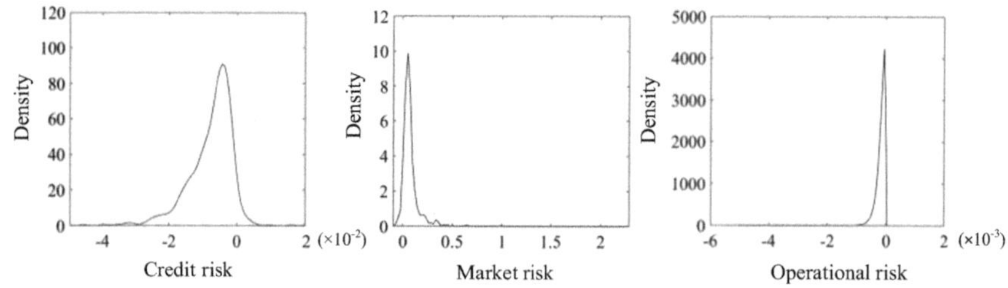

Fig. 6.7 The marginal risk distributions of credit, market, and operational risks

Table 6.9 VaRs of credit, market, and operational risks for Chinese banks

Individual risk (VaR)	95%	98%	99%	99.9%
Credit risk (%)	−1.99	−2.52	−3.14	−3.99
Market risk (%)	−2.06	−2.75	−3.91	−5.08
Operational risk (%)	−0.05	−0.06	−0.07	−0.12

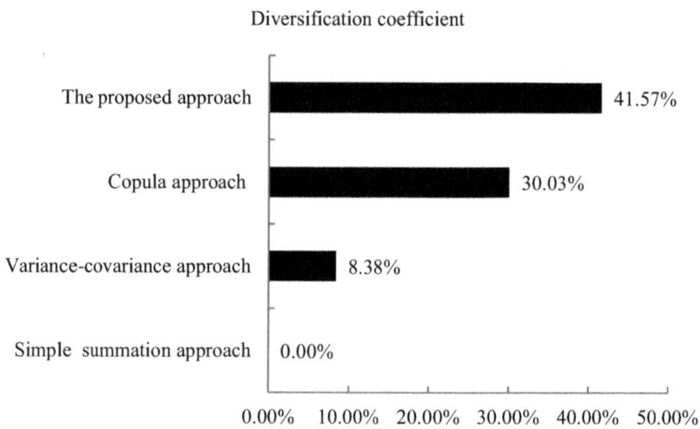

Fig. 6.8 Comparison of diversification benefits of different risk aggregation approaches

It is clear from Table 6.10 that the simple summation approach imposes an upper bound on the total VaR of −9.19% because its completely positive correlation assumption ignores the diversification benefit. The second-largest total risk of −8.42% with the second-smallest diversification coefficient of 8.38% is obtained by the variance–covariance approach, which only considers the linear correlations between risks. The most popular copula approach that considers both the linear and non-linear correlations between risks derives a total risk of −6.43%, which is smaller than the above two approaches. Its diversification coefficient is 30.03%. The proposed approach models the risk correlations in a new and improved way. It produces the smallest total risk of −5.37% and the largest diversification coefficient

Table 6.10 Comparison of total risk VaRs between different risk aggregation approaches

Total risk (VaR)	95%	98%	99%	99.9%
Simple summation approach (%)	−4.10	−5.33	−7.12	−9.19
Variance–covariance approach (%)	−3.75	−4.89	−6.54	−8.42
Copula approach (%)	−3.21	−3.97	−5.09	−6.43
The proposed risk aggregation approach (%)	−2.65	−3.40	−3.86	−5.37

Note The simple summation approach calculates the total VaR by adding each single VaR. The variance–covariance approach uses a matrix of linear correlation coefficients to aggregate individual risks. The Pearson correlation coefficients between credit and market risks, credit and operational risks, and market and operational risks are 0.6652, −0.0012 and 0.0347, respectively. An obvious demerit of the two approaches is that they can only get the total risk VaR but not the entire total risk distribution. For the copula approach, the most appropriate copula is chosen from widely used elliptical copulas and Archimedean copulas by the method proposed by Kojadinovic and Yan (2010). The results show that the t copula with the degree of freedom 5 fits the dependence structure of the three risks best. The fitted correlation coefficients for credit and market risks, credit and operational risks, and market and operational risks are 0.6436, 0.0053 and 0.0062. By using the Monte Carlo simulation, the total risk distribution can be derived and the total risk VaR can be calculated

of 41.57%. This indicates that the three commonly used comparison approaches may overestimate the total risk.

6.5 Conclusions

This chapter proposes a novel two-stage bank risk aggregation approach to aggregate multiple bank risks that are distinctly different from each other. It tactfully divides these risks into two categories based on whether they have common factors or not and then aggregates them by reasonably combing the bottom-up and top-down approaches.

In the empirical analysis, the proposed approach is employed to aggregate the Chinese banking industry's credit, market, and operational risks. The empirical results show that the annual total risk VaR is −5.37% at the 99.9% confidence level. The Chinese banking industry should keep 5676 billion CNY to protect against the potential total losses caused by credit, market, and operational risks. Compared with other top popular approaches, the proposed approach leads to the largest diversification benefit, i.e. 41.57%.

This chapter could be of interest to a financial institution confronted with the problematic issue of multiple risk aggregation in practice. Its introduction of hierarchical copula to the bank risk aggregation domain can also enlighten researchers, regulators, and practitioners.

References

Aas K, Dimakos XK, Oksendal A (2007) Risk capital aggregation. Risk Manage 9(2):82–107. https://doi.org/10.1057/palgrave.rm.8250027

Alessandri P, Drehmann M (2010) An economic capital model integrating credit and interest rate risk in the banking book. J Bank Finance 34(4):730–742. https://doi.org/10.1016/j.jbankfin.2009.06.012

Alexander C, Peizer J (2003) On the aggregation of firm-wide market and credit risk. Henley business school, Reading University. https://doi.org/10.2139/ssrn.478381

Arbenz P, Hummel C, Mainik G (2012) Copula based hierarchical risk aggregation through sample reordering. Insur: Math Econ 51(1):122–133. https://doi.org/10.1016/j.insmatheco.2012.03.009

Baker HK, Filbeck G (2015) Investment risk management. Oxford University Press, New York

Basel Committee on Banking Supervision (2006) International convergence of capital measurement and capital standards: a revised framework. Bank for International Settlements

Basel Committee on Banking Supervision (2010) Developments in modelling risk aggregation. Bank for International Settlements

Begley TA, Purnanandam AK, Zheng K (2016) The strategic under-reporting of bank risk. Rev Financ Stud 30(10):3376–3415. https://doi.org/10.1093/rfs/hhx036

Bellini T (2013) Integrated bank risk modeling: a bottom-up statistical framework. Eur J Oper Res 230(2):385–398. https://doi.org/10.1016/j.ejor.2013.04.031

Bernal O, Gnabo JY, Guilmin G (2014) Assessing the contribution of banks, insurance and other financial services to systemic risk. J Bank Finance 47:270–287. https://doi.org/10.1016/j.jbankfin.2014.05.030

Bernard C, Vanduffel S (2015) A new approach to assessing model risk in high dimensions. J Bank Finance 58:166–178. https://doi.org/10.1016/j.jbankfin.2015.03.007

Biell L, Muller A (2013) Sudden crash or long torture: the timing of market reactions to operational loss events. J Bank Finance 37(7):2628–2638. https://doi.org/10.1016/j.jbankfin.2013.02.022

Bocker K, Hillebrand M (2009) Interaction of market and credit risk: an analysis of inter-risk correlation and risk aggregation. J Risk 11(4):3–29. https://doi.org/10.21314/JOR.2009.198

Brechmann E, Czado C, Paterlini S (2014) Flexible dependence modeling of operational risk losses and its impact on total capital requirements. J Bank Finance 40(1):271–285. https://doi.org/10.1016/j.jbankfin.2013.11.040

Breuer T, Jandacka M, Rheinberger K, Summer M (2010) Does adding up of economic capital for market and credit risk amount always to conservative risk estimates? J Bank Finance 34(4):703–712. https://doi.org/10.1016/j.jbankfin.2009.03.013

Bryant FB, Satorra A (2012) Principles and practice of scaled difference chi-square testing. Struct Equ Modeling 19(3):372–398. https://doi.org/10.1080/10705511.2012.687671

Chapelle A, Crama Y, Hubner G, Peters JP (2008) Practical methods for measuring and managing operational risk in the financial sector: a clinical study. J Bank Finance 32(6):1049–1061. https://doi.org/10.1016/j.jbankfin.2007.09.017

Chen YH, Tu AH (2013) Estimating hedged portfolio value-at-risk using the conditional copula: an illustration of model risk. Int Rev Econ Financ 27:514–528. https://doi.org/10.1016/j.iref.2013.01.006

Côté MP, Genest C (2015) A copula-based risk aggregation model. Can J Stat 43(1):60–81. https://doi.org/10.1002/cjs.11238

Danielsson J, Zigrand JP (2006) On time-scaling of risk and the square-root-of-time rule. J Bank Finance 30(10):2701–2713. https://doi.org/10.1016/j.jbankfin.2005.10.002

Dunis CL, Laws J, Sermpinis G (2011) Higher order and recurrent neural architectures for trading the EUR/USD exchange rate. Quant Finance 11(4):615–629. https://doi.org/10.1080/14697680903386348

Feng J, Li J, Gao L, Hua Z (2012) A combination model for operational risk estimation in a Chinese banking industry case. J Oper Risk 7(2):17–39. https://doi.org/10.21314/JOP.2012.106

Furman E, Wang R, Zitikis R (2017) Gini-type measures of risk and variability: Gini shortfall, capital allocations, and heavy-tailed risks. J Bank Finance 83:70–84. https://doi.org/10.1016/j.jbankfin.2017.06.013

Grundke P (2010) Top-down approaches for integrated risk management: how accurate are they? Eur J Oper Res 203(3):662–672. https://doi.org/10.1016/j.ejor.2009.09.015

Grundke P, Polle S (2012) Crisis and risk dependencies. Eur J Oper Res 223(2):518–528. https://doi.org/10.1016/j.ejor.2012.06.024

Hartmann P (2010) Interaction of market and credit risk. J Bank Finance 34(4):697–702. https://doi.org/10.1016/j.jbankfin.2009.10.013

Hull JC (2012) Risk management and financial institutions, 3rd ed. Hoboken: John Wiley & Sons. https://tainguyenso.vnu.edu.vn/jspui/handle/123456789/51683

Inanoglu H, Jacobs M (2009) Models for risk aggregation and sensitivity analysis: an application to bank economic capital. J Risk Financ Manag 2(1):118–189. https://doi.org/10.3390/jrfm2010118

Jarrow RA, Turnbull SM (2000) The intersection of market and credit risk. J Bank Finance 24(1):271–299. https://doi.org/10.1016/S0378-4266(99)00060-6

Kojadinovic I, Yan J (2010) Modeling multivariate distributions with continuous margins using the copula r package. J Stat Soft 34(9):1–20. https://doi.org/10.18637/jss.v034.i09

Kuritzkes A, Schuermann T (2007) What we know, don't know and can't know about bank risk: a view from the trenches. The known the unknown and the unknowable in financial risk management. Princeton University Press, Princeton

Li J, Zhu X, Lee CF, Wu D, Feng J, Shi Y (2015) On the aggregation of credit, market and operational risks. Rev Quant Financ Acc 44(1):161–189. https://doi.org/10.1007/s11156-013-0426-0

Li J, Zhu X, Chen J, Gao L, Feng J, Wu D, Sun X (2014a) Operational risk aggregation across business lines based on frequency dependence and loss dependence. Math Prob Eng 2014:404208. https://doi.org/10.1155/2014/404208

Li J, Zhu X, Xie Y, Chen J, Gao L, Feng J, Shi W (2014b) The mutual-information based variance-covariance approach: an application to operational risk aggregation in Chinese banking. J Oper Risk 9(3):3–19. https://doi.org/10.21314/JOP.2014.140

Li J, Wei L, Lee CF, Zhu X, Wu D (2017) Financial statements based bank risk aggregation. Rev Quant Finance Account 1–22. https://doi.org/10.1007/s11156-017-0642-0

Li J, Li J, Zhu X, Yao Y, Casu B (2020) Risk spillovers between FinTech and traditional financial institutions: evidence from the U.S. Int Rev Financ Anal. Online. https://doi.org/10.1016/j.irfa.2020.101544

Okhrin O, Tetereva A (2017) The realized hierarchical Archimedean copula in risk modeling. Econometrics 5(2):26. https://doi.org/10.3390/econometrics5020026

Peng W, Hu S, Chen W, Zeng YF, Yang L (2019) Modeling the joint dynamic value at risk of the volatility index, oil price, and exchange rate. Int Rev Econ Financ 59:137–149. https://doi.org/10.1016/j.iref.2018.08.014

Rosenberg JV, Schuermann TA (2006) General approach to integrated risk management with skewed, fat-tailed risks. J Financ Econ 79(3):569–614. https://doi.org/10.1016/j.jfineco.2005.03.001

Wei Y, Wang Y, Huang D (2010) Forecasting crude oil market volatility: further evidence using GARCH-class models. Energy Econ 32(6):1477–1484. https://doi.org/10.1016/j.eneco.2010.07.009

Wei L, Li J, Zhu X (2018) A factor-integral approach to aggregating credit and market risks. J Risk Forthcoming

Wei L, Li G, Zhu X, Li J (2019) Discovering bank risk factors from financial statements based on a new semi-supervised text mining algorithm. Accounting & Finance, 59(3), 1519–1552. https://doi.org/10.1111/acfi.12453

Yang L, Cai XJ, Li M, Hamori S (2015) Modeling dependence structures among international stock markets: evidence from hierarchical Archimedean copulas. Econ Model 51:308–314. https://doi.org/10.1016/j.econmod.2015.08.017

Zhu X, Li J, Wu D (2020) Simultaneously capturing multiple dependence features in bank risk aggregation: A mixture copula framework. In: Lee CF, Lee JC (eds) Handbook of financial econometrics, mathematics, statistics, and technology. World Scientific, New Jersey

Chapter 7
Bank Risk Aggregation Based on Income Statement and Balance Sheet

7.1 Introduction

One major challenge in risk aggregation is the risk data used for establishing marginal risk distributions (Basel Committee on Bank Supervision, BCBS for short 2003a, b). Many previous studies have attempted to use simulated risk data to measure credit risk, market risk and liquidity risk (Dimakos and Aas 2004; Acerbi and Scandolo 2008), which can hardly replace the real data. For the operational risk, external real data are often used to supplement insufficient internal loss data. However, some remain skeptical of the external operational risk data (BCBS 2003a, b; Chavez et al. 2006). Thus, the shortage and inconsistency of risk data limit the reliability and validity of risk aggregation results.

Recent research has, instead, used publicly available industry-wide data from a set of commercial banks' financial statements to develop empirical proxies for different risk types. Although financial statements data have some drawbacks, such as lower reporting frequency (usually published quarterly), different accounting standards across the world (Bae et al. 2008) and poor accounting quality (Saito 2012), collecting risk data from financial statements is still a satisfactory way to resolve the problems of data shortage and data inconsistency.

Some have attempted to aggregate marginal risks based on-balance sheet data. Kretzschmar et al. (2010) implement a fully-integrated risk analysis based on-balance sheet asset positions. Mapping profit and loss (P&L) items from income statement into risk types is another feasible way to obtain risk data. As researchers have realized that risk is defined in terms of earnings volatility (Rajan 2006), P&L items from income statement that are created by earnings volatility can be used as proxies for risks (Kuritzkes and Schuermann 2007). Thus, Kuritzkes and Schuermann (2007) get risk P&L successfully by mapping income statement items of US banks into risk types. Given the significant accounting difference between income statements in the US and China, Li et al. (2012) use data of risk P&L to measure Chinese banks' risks by establishing a mapping relationship between Chinese banks' income statements and risk types.

J. Li et al., *Financial Statements-Based Bank Risk Aggregation*, Innovation in Risk Analysis, https://doi.org/10.1007/978-981-19-0408-0_7

The above studies merely focus on one piece of financial statements, either income statement or balance sheet, while Inanoglu and Jacobs (2009) match risk types with items from both income statement and balance sheet. In particular, the liquidity risk is mapped into balance sheet items and the credit, market and operational risks are mapped into income statement items. However, this correspondence creates a problem of discrepancies in attributes of proxies for different risk types. By contrast, Rosenberg and Schuermann map risk types into income statement and balance sheet to obtain risk P&L and risk exposure, respectively. By doing so, they collect data from both income statement and balance sheet simultaneously.

To summarize, previous studies have suggested a relatively complete risk aggregation framework based on financial statements by mapping balance sheet and income statement items into multiple risk types. Thus, in this chapter, we adopt the bank risk aggregation approach based on income statements and balance sheets to aggregate the credit, market, liquidity and operational risks of the Chinese banking sector. In the empirical analysis, based on 16 early Chinese listed commercial banks spanning the period 2007–2018, we collect 743 pieces of risk data from financial statements to aggregate credit, market, liquidity and operational risks of the Chinese banking sector. Furthermore, we analyze the difference of bank risks in banks of different sizes by constructing a large, general and small Chinese commercial bank.

The rest of this chapter is structured as follows. Section 7.2 illustrates the bank risk aggregation approach based on income statement and balance sheet. Section 7.3 describes the process of data collection. In Sect. 7.4, the approach is applied to aggregating the credit, market, liquidity and operational risks of the Chinese banking sector. Section 7.5 summarizes the conclusions.

7.2 The Mapping Approach of Balance Sheet and Income Statement

7.2.1 Mapping Income Statement Items into Risk Profits & Losses

To obtain risk data from financial statements, we map income statement items and balance sheet assets into risk types.

The more complete and reasonable mapping relationship between financial statements and risk types is shown in Table 7.1. By mapping income statement items and balance sheet assets into risk types, we obtain risk P&L and risk exposure, respectively. Although these mappings are hardly perfect, we believe they still provide a reasonable approximation of risk type attribution.

As shown in Table 7.1, credit risk P&L items related to credit risk exposure are net interest income and loan impairment loss. The reason is that changes in interest-bearing assets will lead to changes in net interest income, and loan impairment loss should be recorded if there is any indication that loans have suffered an impairment

Table 7.1 The correspondence between risk types and income statement and balance sheet

Risk type	Income statement	Balance sheet
Credit risk	Net interest income – Loan impairment loss	Interest-bearing assets – Loan loss provisions
Market risk	Gains or losses from changes in fair values of financial instruments +Net foreign exchange differences	Traded financial assets +Investment real estate +Derivatives +Precious metals
Liquidity risk	Net investment income – Investment income from associates and joint ventures	Traded financial assets +Investment real estate +Derivatives +Precious metals +Held-to-maturity investments +Financial assets available for sale
Operational risk	Fees and commissions income – Other assets impairment loss – Business tax and surcharges – Operation and administrative expense +Other business income +Net non-operating income	Total assets

loss (Kwak et al. 2009). Thus, credit risk P&L is equal to net interest income minus loan impairment loss.

Risk P&L items related to market risk exposure are gains or losses from fair values of financial instruments and net foreign exchange differences. The reason is that gains or losses from fair values of financial instruments are affected by fluctuations in prices of financial instruments, and net foreign exchange differences are determined by changes in foreign exchange. Therefore, the sum of these two accounts is a proxy for market risk P&L.

Investment income from associates and joint ventures is generated by long-term equity investments, which is made to control or influence other companies, not to get short-term investment income. Thus, liquidity risk P&L equals net investment income minus investment income from associates and joint ventures.

The remaining P&L items in the income statement serve as a proxy for operational risk P&L because operational risk is a typical non-financial risk and represents the volatility of residual earnings, which cannot be categorized into market, credit or liquidity risk.

7.2.2 *Mapping Balance Sheet Items into Risk Expsoures*

Credit risk exposure is equal to interest-bearing assets minus loan loss provisions. Interest-bearing assets include loans, due from the central bank, due from banks and other financial institutions, accounts receivable investment, buying back the sale of financial assets, lending to banks and other financial institutions and bonds.

Regarding market risk exposure, it includes traded financial assets, investment real estate, derivatives and precious metals because all these assets are influenced by market factors (i.e. price, interest rate, foreign exchange).

Liquidity risk exposure includes traded financial assets, investment real estate, derivatives, precious metals, held-to-maturity investments and financial assets available for sale. Net investment income reflects the gains or losses from trading ready to liquidate financial assets.

For operational risk, Rosenberg and Schuermann (2006) deem that all assets and activities of the bank are in some way subject to operational risk. We follow this standpoint that operational risk exposure consists of total on-balance and off-balance sheet assets.

7.2.3 Procedure of Risk Measurement and Aggregation

Risk P&L items from income statements are not comparable among banks as they differ in scale, capital allocation, investment strategy, and management level (Rosenberg and Schuermann 2006). To allow direct comparison across banks, risk P&L needs to be converted into a "risk return" based measure. Following Kretzschmar et al. (2010), we use the data preprocessing method to obtain a specific bank's risk return. The procedure of data preprocessing can be divided into the following three steps:

Firstly, we convert risk P&L into a "risk return" based measure. Since risk P&L is generated by assets exposed to risk, an obvious approach for doing this would be to divide risk P&L by assets to yield a return on assets measure. In this chapter, bank assets are defined as risk exposures. Thus, the risk return is the ratio of risk P&L to risk exposure. We then define the marginal risk return for the *i*th bank, *j*th risk in period *t* as

$$r_{i,j,t} = \left(\frac{R_{i,j,t}}{RE_{i,j,t}}\right) \tag{7.1}$$

where $r_{i,j,t}$, $R_{i,j,t}$ and $RE_{i,j,t}$ stand for the risk return, risk P&L and risk exposure of bank *i*, risk *j* in period *t*, respectively.

In the second step, we compute the expected risk return and deviation from risk return. The risk return can be divided into two parts: the expected risk return and deviation from risk return. The expected risk return for a bank is the average risk return over the sample period, which reflects the bank's characteristic in terms of scale, capital allocation, investment strategy and management level. The deviation from risk return is computed by subtracting the average risk return over the sample period (expected risk return) for each bank, reflecting the macroeconomic background and operating conditions of the whole banking industry. Thus, a bank's risk return is determined by both market and individual information. Specifically, the expected risk return is defined as

$$\overline{r}_{i,j} = \left(\frac{1}{T_i} \sum_{t=1}^{Ti} r_{i,j,t} \right) \tag{7.2}$$

and the deviation from risk return is defined as

$$\Delta_{i,j,t} = r_{i,j,t} - \overline{r}_{i,j} \tag{7.3}$$

where bank i is observed for T_i periods. $\overline{r}_{i,j}$ denotes the expected risk return for bank i and risk j over the sample T_i period. $\Delta_{i,j,t}$ denotes the deviation from risk return of bank i, risk j in period t.

Finally, we obtain a typical bank's risk returns to model marginal risk distributions. For a typical bank, its risk return is determined by market information and individual information. The market information is composed of all sample banks' deviation from risk return. Thus, by combining all sample banks' deviation from risk return and the typical bank's ($i = k$) expected risk return, we finally compute a typical bank's risk return. Specifically, the typical bank's risk return is written as

$$r_{k,j,t} = \overline{r}_{k,j} + \Delta_{j,t} = \overline{r}_{k,j} + \sum_{i} \Delta_{i,j,t} \tag{7.4}$$

where $r_{k,j,t}$ is the risk return of bank k, risk j in period t. $\overline{r}_{k,j}$ stands for the expected risk return of bank k and risk j. $\Delta_{j,t}$ denotes the summation of deviation from risk return of risk j in period t of all sample banks, reflecting the market information of risk j in period t.

Value-at-Risk (VaR), which has become a standard model for measuring and assessing risk, is used to measure marginal risk in this chapter (Huang 2013; Hsu et al. 2012). VaR is defined as a quantile of the distribution of risk returns. Thus, the larger negative value or smaller positive value of VaR corresponds to the higher level of risk (Rosenberg and Schuermann 2006). To single aggregate VaRs into total risk, we adopt the simple summation approach, which is one of the most basic and widely used risk aggregation approaches. Some risk aggregation approaches have emerged so far. To illustrate, simple summation, var-covar and copula approaches are the three main risk aggregation approaches. All of them have strengths and weaknesses (Li et al. 2015).

Specifically, the simple summation approach is one of the most basic and widely used approaches to aggregate risk (Rosenberg and Schuermann 2006; Inanoglu and Jacobs 2009; Kretzschmar et al. 2010). It has several features. One is that it is the briefest approach that calculates total risk by just adding stand-alone risks. Another is that it is found to be more conservative compared with other risk aggregation approaches (Embrechts et al. 1999). Such an approach implicitly assumes that all risks are perfectly correlated. That is to say, great losses occur simultaneously, which imposes an upper bound on the true total risk (Dimakos and Aas 2004). Therefore, many papers use the simple summation approach to aggregate marginal risks, such as Rosenberg and Schuermann (2006), Inanoglu and Jacobs (2009) and Kretzschmar et al. (2010). Given that this chapter aims to analyze the impact of OBS activities on total risk rather than risk aggregation approaches, we adopt the widely used simple summation approach to aggregate different risk types.

7.3 Data Description

As discussed in previous chapters, to analyze Chinese banking sector, we collected quarterly panel data over the period 2007–2018 from early 16 A-share listed Chinese commercial banks (Table 4.1) to ensure consistency of accounts. The quarterly data of ABC and CEB from 2007 to 2009 are unavailable because they were listed in 2010. Besides, 2007-Q2 data of BOBJ, 2007-Q1 data of BONJ, BONB and CCB are also missing. Getting rid of these exceptional cases, we finally obtained 743 pieces of valid data to model individual risk distributions.

Our empirical analysis is based on quarterly data while loan impairment loss and loan loss provision are disclosed only in annual and semi-annual financial reports. Hence, we need to make simple assumptions to obtain quarterly data of these accounts. As for loan impairment loss, which is part of assets impairment loss, can be calculated based on known quarterly assets impairment loss. Specifically, we first calculate R, which is a ratio of loan impairment loss to assets impairment loss based on annual and semi-annual data. Then, we calculate the mean value of this ratio over the sample period $(\overline{R})$. Herein, we make a simple assumption that the quarterly R is equal to $\overline{R}$. Thus, the quarterly loan impairment loss is obtained by multiplying $\overline{R}$ and quarterly assets impairment loss. Likewise, the quarterly loan loss provision, which is determined by the quality of loans, can be obtained based on the known quarterly loans. Specifically, we define R' as the ratio of loan loss provision to loans and $\overline{R'}$ as the mean value of R' over the sample period. Therefore, the quarterly loan loss provision is obtained by multiplying $\overline{R'}$ with quarterly loans based on the assumption that the quarterly R' is equal to $\overline{R'}$.

Among the sample of all 16 early-listed Chinese commercial banks from 2007 to 2018, the number of financial statements data for a single bank is up to 48, which is too small to perform the empirical analysis. Thus, in order to address the problem of data shortage and provide empirical insights into the total risk of Chinese commercial banks, we construct hypothetical banks following Rosenberg and Schuermann (2006), Kretzschmar et al. (2010) and Alessandri and Drehmann (2010).

In particular, we use median assets to characterize hypothetical banks, and then a hypothetical general bank, a large hypothetical bank and a small hypothetical bank are constructed for comparison. "The big four" stated-owned banks are the four largest banks by assets in Chinese banking system. However, ABC went public relatively late so that the amount of financial statements data is relatively smaller. Therefore, the asset size of the large hypothetical bank is the average of the rest three state-owned banks (ICBC, BOC and CCB). Correspondingly, the asset size of the small hypothetical bank is the average of the three smallest banks by assets (BOBJ, BONB and BONJ). The hypothetical general bank is constricted by averaging the 16 listed banks. Using this median approach to constructing hypothetical banks, at the end of 2018, the bank sizes in terms of risk-weighted assets for the general, large and small hypothetical banks are 14,564, 6041 and 1161 billion CNY, respectively.

For either of these three hypothetical banks, the amount of data is 743, which is much larger than that of a real-world bank. Furthermore, the hypothetical banks constructed by us capture the characteristics of real-world banks' asset sizes, so they are the typical banks in Chinese banking system. In a word, performing empirical analysis based on typical hypothetical banks not only addresses the problem of data shortage but also achieves general conclusions.

7.4 Empirical Results

7.4.1 Marginal Risk Results of the Chinese Commercial Bank

The marginal risk distribution is decided by the deviation from risk return and the expected risk return by referring to Eq. (7.4). The deviation from risk return, which reflects the macroeconomic background and operating conditions of the banking industry, decides the shape of marginal risk distribution. The expected risk return that reflects a bank's features decides the horizontal axis coordinates of the marginal risk distribution. The characteristics of deviation from risk return are presented numerically in Table 7.2, and the shapes of marginal risk distributions are visually shown in Fig. 7.1.

In terms of return on risk exposure, market risk has the highest volatility (14.48%). Liquidity risk has the fattest tails (kurtosis = 173.65). The volatility (2.48%) of liquidity and kurtosis (46.26%) of market risk come in second. The negative kurtosis present in credit risk (−0.83) and the relatively lower positive kurtosis for operational risk (8.71) suggest that credit risk and operational risk have thinner tails, with volatilities of credit and operational risks being 1.21% and 0.82%, respectively.

Table 7.2 Descriptive statistics for four marginal risks of three hypothetical banks

Statistics	Credit risk	Market risk	Liquidity risk	Operational risk
σ (%)	1.21	14.48	2.48	0.82
Skewness	0.34	2.98	11.29	−2.46
Kurtosis	−0.83	46.26	173.65	8.71

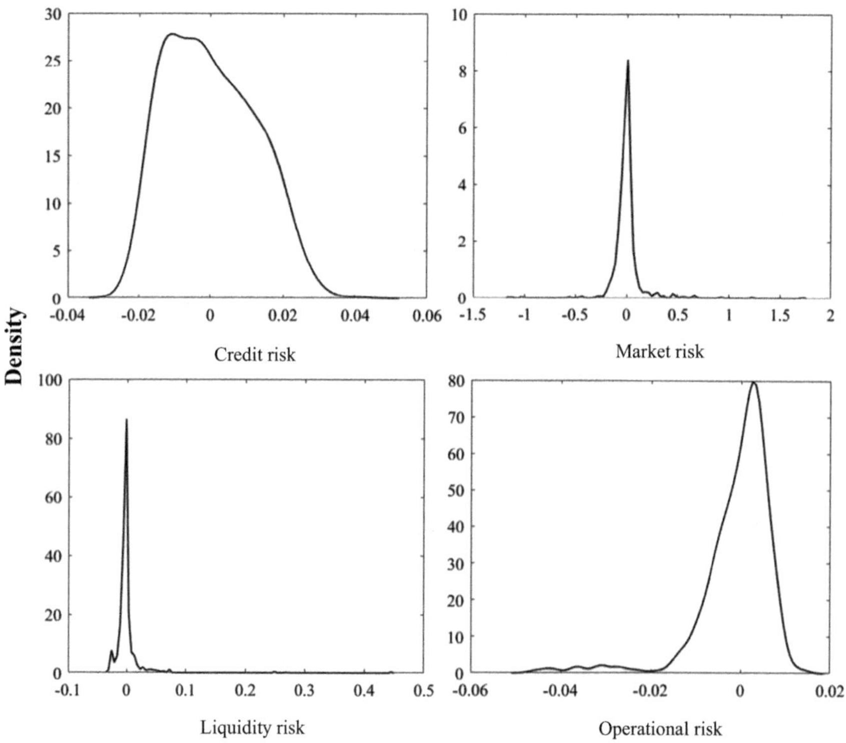

Fig. 7.1 Marginal distributions of credit, market, liquidity and operational risks

The shape of operational risk distribution is left-skewed at −2.46, while the other three risk types are right-skewed. Specifically, credit risk is moderately right-skewed at 0.34, market risk is right-skewed at 2.98 and liquidity risk is more significantly right-skewed at 11.29.

Table 7.3 shows VaR values of individual risks at different confidence levels. The larger negative value or smaller positive value of VaR corresponds to the higher level of risk. At 99.9% confidence level, market risk is the highest of the four types of risk for all three banks. The market risk VaR values of the large, general and small banks are −24.70%, −13.62% and −12.11%, respectively. Furthermore, we found that the market risk declined with the decrease of bank size. Thus, the small bank has better market risk management ability. For all the three banks, operational risk is the second highest risk among the four types of risks. The VaR values of the operational risk of the large, general and small banks are −2.10%, −5.13% and −1.08%, respectively. We can see that the large and small banks have better operational risk management ability while the general bank's operational risk is relatively higher. Then is the liquidity risk, whose VaR values are −2.76%, −2.71% and −0.16% of the large, general and small banks, respectively, indicating that the small bank's liquidity risk is relatively lower. The VaR values of credit risk for the large, general and small bank

Table 7.3 Four marginal risks VaR of three hypothetical banks at different confidence levels

	Confidence level	99.9%	99%	98%	95%
Large bank	Credit risk (%)	0.48	0.68	0.75	0.85
	Market risk (%)	−24.70	−19.99	−19.17	−13.28
	Liquidity risk (%)	−2.76	−0.66	−0.65	−0.39
	Operational risk (%)	−2.10	−1.75	−1.75	−1.12
General bank	Credit risk (%)	0.36	0.57	0.68	0.85
	Market risk (%)	−13.62	−27.40	−17.93	−13.23
	Liquidity risk (%)	−2.71	−2.28	−2.17	−1.60
	Operational risk (%)	−5.13	−4.64	−3.63	−1.86
Small bank	Credit risk (%)	0.44	0.68	0.74	0.89
	Market risk (%)	−12.11	−11.28	−8.40	−6.31
	Liquidity risk (%)	−0.16	−0.15	−0.13	−0.01
	Operational risk (%)	−1.80	−1.72	−1.71	−1.28

are 0.48%, 0.36% and 0.44%, respectively. Thus, the credit risk is the lowest among the four types of marginal risks for all three banks.

7.4.2 Total Risk Results of the Chinese Commercial Bank

After getting VaR values of marginal risks, we then calculate the total risk by adding single VaRs. Add-VaR, the total risk in terms of return as a percent of total risk exposure, is the simple weighted summation of marginal risks. In 2018, the large bank's marginal risk weights are 40.49%, 3.19%, 3.19% and 53.14% for credit, market, liquidity and operational risks, respectively. As for the general bank, the marginal risk weights for credit, market, liquidity and operational risks are 40.02%, 3.52%, 4.41% and 52.05%, respectively. Regarding the small bank credit, market, liquidity and operational risk weights are 41.49%, 3.29%, 12.05% and 43.16%, respectively. We can see that for all the three banks, the risk weight of operational risk is the highest, followed by the credit risk. The risk weight of liquidity risk ranks third and the risk weight of market risk is the lowest. Furthermore, the liquidity risk weight of the small bank is significantly higher than that of the large or the general bank, which indicates that the proportion of assets bearing liquidity risk is higher in the asset portfolio of the small bank.

Then, the total loss, which is the total risk in terms of losses is calculated. For simplicity in what follows, we shall refer to both Add-VaR and total loss as the total risk. Our quarterly financial statements data enable a quarterly view of total risk while the typical horizon of losses is one year. To transform the total quarterly loss into the annual total loss, we apply the *square-root-of-time rule,* which is commonly used to scale an estimated quantile of a return distribution to a lower frequency T by

Table 7.4 2018 Add-VaRs and total losses of the three hypothetical banks

Confidence level	99.9%	99%	98%	95%
	Large hypothetical bank			
Total risk (Add-VaR) (%)	−3.51	−3.03	−2.85	−2.34
Total risk (losses in billion CNY)	1591.11	1370.87	1290.83	1057.85
	General hypothetical bank			
Total risk (Add-VaR) (%)	−7.36	−3.67	−3.03	−2.82
Total risk (losses in billion CNY)	1338.71	667.80	550.49	513.68
	Small hypothetical bank			
Total risk (Add-VaR) (%)	−2.50	−2.34	−2.06	−1.88
Total risk (losses in billion CNY)	95.40	88.98	78.33	71.61

the multiplication of $\sqrt{T}$ (Danielsson and Zigrand 2006). Table 7.4 gives a summary overview of the two hypothetical banks' total risks in 2018.

As shown in Table 7.4, there is no significant difference between the total risk VaR values of these three banks, while the total loss of the large commercial bank is bigger due to its larger scale. Specifically, at 99.9% confidence level, 2018 annual Add-VaR values of the large, general and small banks are −3.51%, −7.36% and −2.50%, respectively. The large bank may suffer an annual total loss of 1159 billion CNY and the annual loss of the general bank is 1339 billion CNY while the small bank's annual total loss is equal to 95 billion CNY in 2018.

7.5 Conclusions

In this chapter, we aggregate bank risks using the bank risk aggregation approach based on balance sheet and income statement. By mapping different risk types into balance sheet and income statement, we can obtain the data of risk exposure and risk P&L. In the empirical analysis, we apply this approach to aggregate credit, market, liquidity and operational risks by using a sample of all 16 Chinese early-listed commercial banks over 2007–2018. Then we empirically study the difference among risks faced by different size banks by constructing three hypothetical Chinese commercial banks.

Our empirical results show that for all hypothetical banks, the market risk is the highest. Operational risk comes second, followed by liquidity risk. The credit risk is the lowest. At 99.9% confidence level, 2018 annual Add-VaR values of the large, general and small banks are −3.51%, −7.36% and −2.50%, respectively. The large bank may suffer an annual total loss of 1159 billion CNY, and the annual loss of the general bank is 1339 billion CNY. While the small bank's annual total loss is equal to 95 billion CNY in 2018.

Nevertheless, this chapter has several limitations. The correspondence between risk types and financial statements is kind of rough. For instance, net interest income bears credit risk and market risk simultaneously. Moreover, whether all assets are subject to operational risk is still open to question. In future studies, the employment of other information may help calibrate the corresponding relationship to some extent.

References

Acerbi C, Scandolo G (2008) Liquidity risk theory and coherent measures of risk. Quant Financ 8(7):681–692. https://doi.org/10.1080/14697680802373975

Alessandri P, Drehmann M (2010) An economic capital model integrating credit and interest rate risk in the banking book. J Bank Finance 34(4):730–742. https://doi.org/10.1016/j.jbankfin.2009.06.012

Bae KH, Tan H, Welker M (2008) International GAAP differences: the impact on foreign analysts. Account Rev 83(3):593–628. https://doi.org/10.2308/accr.2008.83.3.593

Basel Committee on Banking Supervision (2003a) The 2002 loss data collection exercise for operational risk: summary of the data collected. Bank for International Settlements, Basel, Switzerland

Basel Committee on Banking Supervision (2003b) Trends in risk integration and aggregation. Bank for International Settlements, Basel, Switzerland

Chavez DV, Embrechts P, Nešlehová J (2006) Quantitative models for operational risk: extremes, dependence and aggregation. J Bank Finance 30(10):2635–2658. https://doi.org/10.1016/j.jbankfin.2005.11.008

Dimakos XK, Aas K (2004) Integrated risk modelling. Stat Model 4(4):265–277. https://doi.org/10.1191/1471082X04st079oa

Embrechts P, McNeil A, Straumann D (1999) Correlation: pitfalls and alternatives. Risk 12(5):69–71

Hsu CP, Huang CW, Chiou W (2012) Effectiveness of copula-extreme value theory in estimating value-at-risk: empirical evidence from Asian emerging markets. Rev Quant Financ Acc 39(4):447–468. https://doi.org/10.1007/s11156-011-0261-0

Huang A (2013) Value at risk estimation by quantile regression and kernel estimator. Rev Quant Financ Acc 41(2):225–251

Inanoglu H, Jacobs M (2009) Models for risk aggregation and sensitivity analysis: an application to bank economic capital. J Risk Financ Manag 2(1):118–189. https://doi.org/10.3390/jrfm2010118

Kretzschmar G, McNeil AJ, Kirchner A (2010) Integrated models of capital adequacy-why banks are undercapitalized. J Bank Finance 34(12):2838–2850. https://doi.org/10.1016/j.jbankfin.2010.02.028

Kuritzkes A, Schuermann T (2007) What we know, don't know and can't know about bank risk: a view from the trenches. In: Diebod FX et al (eds) The known, the unknown and the unknowable in financial risk management. Princeton University Press, Princeton

Kwak W, Lee HY, Eldridge SW (2009) Earnings management by Japanese bank managers using discretionary loan loss provisions. Rev Pac Basin Financ Mark Policies 12(01):1–26. https://doi.org/10.1142/S0219091509001526

Li J, Feng J, Sun X, Li M (2012) Risk integration mechanisms and approaches in banking industry. Int J Inf Technol Decis Mak 11(6):1183–1213. https://doi.org/10.1142/S0219622012500320

Li J, Zhu X, Lee CF, Wu D, Feng J, Shi Y (2015) On the aggregation of credit. Market Inance Account 44(1):161–189. https://doi.org/10.1007/s11156-013-0426-0

Rajan RG (2006) Has finance made the world riskier? Eur Financ Manag 12(4):499–533. https://doi.org/10.1111/j.1468-036X.2006.00330.x

Rosenberg JV, Schuermann TA (2006) General approach to integrated risk management with skewed, fat-tailed risks. J Financ Econ 79(3):569–614. https://doi.org/10.1016/j.jfineco.2005.03.001

Saito Y (2012) The demand for accounting information: young NASDAQ listings versus S&P 500 NYSE listings. Rev Quant Financ Acc 38(2):149–175. https://doi.org/10.1007/s11156-010-0223-y

Chapter 8
Bank Risk Aggregation with Off-Balance Sheet Items

8.1 Introduction

Some characteristics of off-balance sheet (OBS) activities, such as blind expansion and high risk, made the existence of OBS activities a key factor that caused destabilization during the subprime crisis (Brunnermeier 2009). Basel II, however, was widely seen as having failed to adequately capture the risks posed by OBS activities (Acharya and Richardson 2009; Blundell-Wignall and Atkinson 2010). Essentially, OBS risk should be regarded as an indispensable part of a bank's overall risk because both on- and off-balance sheet activities lead to bank risks (BCBS 1986). Basel Committee has already made great strides in strengthening the regulatory capital framework to cover risks, whatever the source (BCBS 2010). Thus, a reliable risk aggregation model is urgently needed to capture both on- and off-balance sheet risks.

One major challenge in risk aggregation is the risk data used for establishing marginal risk distributions (BCBS 2003). Many previous studies have attempted to use simulated risk data to measure credit risk, market risk and liquidity risk (Dimakos and Aas 2004; Acerbi and Scandolo 2008), which can hardly replace the real data. For the operational risk, external real data are often used to supplement insufficient internal loss data. However, some remain skeptical of the external operational risk data (BCBS 2003; Chavez-Demoulin et al. 2006). Therefore, the shortage and inconsistency of risk data limit the reliability and validity of risk aggregation results.

Recent researches have, instead, used publicly available industry-wide data from a set of commercial banks' financial statements to develop empirical proxies for different risk types. Although financial statements data have some drawbacks, such as lower reporting frequency (usually published quarterly), different accounting standards across the world (Bae et al. 2008) and poor accounting quality (Saito 2012), collecting risk data from financial statements is still a satisfactory way to resolve the problems of data shortage and data inconsistency.

J. Li et al., *Financial Statements-Based Bank Risk Aggregation*, Innovation in Risk Analysis, https://doi.org/10.1007/978-981-19-0408-0_8

Some have attempted to aggregate marginal risks based on-balance sheet data. Kretzschmar et al. (2010) implement a fully-integrated risk analysis based on-balance sheet asset positions. However, the exclusion of OBS derivatives from asset portfolios weakens the effectiveness of qualitative conclusions. Given the importance of OBS items, Drehmann et al. (2010) not only take account of balance sheet assets and liabilities, which have been considered by Alessandri and Drehmann (2010) for integrating credit and interest rate risks, but also pay attention to OBS items. Such a modification makes the hypothetical bank reflect a real commercial bank more accurately.

Mapping profit and loss (P&L) items from income statements into risk types is another feasible way to obtain risk data. As researchers have realized that risk is defined in terms of earnings volatility (Rajan 2006), P&L items from income statements that are created by earnings volatility can be used as proxies for risks (Kuritzkes and Schuermann 2007). Thus, Kuritzkes and Schuermann (2007) get risk P&L successfully by mapping income statement items of U.S. banks into risk types. Given the significant accounting difference between income statements in the U.S. and China, Li et al. (2012) use data of risk P&L to measure Chinese banks' risks by establishing a mapping relationship between Chinese banks' income statements and risk types.

The above studies merely focus on one piece of financial statements, either income statement or balance sheet, while Inanoglu and Jacobs (2009) match risk types with items from both income statements and balance sheets. In particular, the liquidity risk is mapped into balance sheet items, the credit, market and operational risks are mapped into income statement items. However, this correspondence creates a problem of discrepancies in attributes of proxies for different risk types. By contrast, Rosenberg and Schuermann map risk types into income statements and balance sheets to obtain risk P&L and risk exposure, respectively. By doing so, they collect data from both income statements and balance sheets simultaneously. Although Rosenberg and Schuermann (2006) realized that OBS items can be larger and the results may be somewhat arbitrary because only on-balance sheet items are considered, they still followed the usual practice of ignoring OBS items.

To summarize, previous studies have suggested a relative complete risk aggregation framework based on financial statements by mapping balance sheet and income statement items into multiple risk types. Nevertheless, ignoring OBS items is regarded as the usual practice in risk aggregation, which may lead to deviations in conclusions because both on-balance and off-balance sheet assets are exposures to risk in the context of the generation of risk P&L items.

Since the 1980s, the product assortment of commercial banks has shifted sharply from traditional on-balance sheet activities to non-traditional OBS activities because of the tendency to avoid supervision and pursue higher yield in the midst of increasingly intense competition (Boyd and Gertle 1994). With the rapid expansion of OBS activities, they have become one of the main pillars of banks. According to the *China Financial Stability Report 2019*, at the end of 2018, OBS items exceeded 340 trillion CNY, accounting for 125% of total on-balance sheet assets. This suggests that as the burgeoning banking business, OBS activities have reached an important stage in the Chinese banking sector (Hou et al. 2015).

However, OBS activities trigger additional risks while bringing considerable income, and the role of OBS items in systemic vulnerability was highlighted during the subprime crisis. As early as 1988, the business scope under supervision had already extended from balance sheet items to OBS items (BCBS 1988). The China banking regulatory commission (CBRC) also published a policy document titled *Risk Management Guidelines of Commercial Banks' off-balance Sheet Business* to regulate OBS activities in 2011.

In this chapter, therefore, we improve the financial statements based risk aggregation framework by mapping OBS items into risk types to get more accurate and rational risk distributions. In the experiment, we construct two hypothetical banks of different sizes for comparison as the expansion of OBS activities is linked to bank size (DeYoung and Rice 2004). Through a dataset that covers all 16 Chinese early-listed commercial banks spanning the period 2007–2018, we aggregate credit, market, liquidity and operational risks. Then by comparing the total risk with and without OBS activities, we empirically prove that OBS activities indeed affect total risk, and the impact depends on bank size. Thus, ignoring OBS activities will lead to deviations in risk aggregation results. Furthermore, we analyze how the subprime crisis affects Chinese commercial banks' risks by dividing the sample into during and after the subprime crisis.

The remainder of this chapter is organized as follows. The next section presents the improved financial statements based risk aggregation framework in detail. Section 8.3 describes data collection and preprocessing procedures. Section 8.4 discusses the major empirical results. Section 8.5 concludes with a summary of findings, limitations and future research directions.

8.2 The Mapping Approach Between OBS and Risk Types

8.2.1 The Mapping Relationship Between Risk Types and OBS Items

Liquidity dried up during the subprime crisis, so liquidity risk is a challenge to a bank in times of stress (Cornett et al. 2011). Hence, Basel III not only requires sound credit, market and operational risks management in pillar one standards but also enhances liquidity risk supervision in pillar two requirements (BCBS 2010). In line with Basel III, we intend to aggregate credit, market, liquidity and operational risk in this chapter.

The major risk for most banks is the credit risk, which is resulted from the counterparty failure (BCBS 1988; Mustika et al. 2015). The risk of loss arising from adverse price movements in a bank's principal trading positions is referred to as market risk (BCBS 1996). Liquidity risk occurs when a bank fails to fund increases in assets or meet obligations as they become due without incurring unacceptable losses (BCBS 2008). A widely used definition of operational risk is the loss resulting from inadequate or failed internal processes, people and systems, or external events (BCBS 2006; Li et al. 2014).

To obtain risk data from financial statements, we map income statement items and on-balance and off-balance sheet assets into risk types. Compared with the existing financial statements based risk aggregation framework, in which OBS assets are ignored, we not only establish the mapping relationship between on-balance sheet assets and risk types but also map OBS assets into risk types. Essentially, OBS risk is an indispensable part of a bank's overall risk because both on- and off-balance sheet activities create bank risks (BCBS 1986). Furthermore, the fast-growing of OBS activities makes the scale of OBS items is too large to ignore (Karim et al. 2013; Hou et al. 2015). Thus, the incremental information contained in OBS accounting disclosures (Seow and Tam 2002) makes the mapping relationship proposed by us more complete and reasonable.

Then, we identify OBS items that will be incorporated into our improved risk aggregation framework. The definition of OBS activities in a narrow sense consists of commitments, guarantees, derivatives and investment banking business (BCBS 1988). In China, however, OBS financial derivatives are accounted for at fair value in the balance sheet from 2007 onwards as per the new accounting standards. In this context, the *Risk Management Guidelines of Commercial Banks' off-balance Sheet Business* issued in 2011 by CBRC divided OBS business into guarantee business and commitment business. Unfortunately, the disclosure of OBS items is limited and varies from bank to bank. Hence, we just take part of OBS items into risk aggregation, including credit commitment, capital expenditure commitment and operating lease commitment.

The more complete and reasonable mapping relationship between financial statements and risk types is shown in Fig. 8.1. By mapping income statement items and on-balance and off-balance sheet assets into risk types, we obtain risk P&L and risk exposure, respectively. Although these mappings are hardly perfect, we believe they still provide a reasonable approximation of risk type attribution.

As shown in Fig. 8.1, credit risk exposure equals interest-bearing assets minus loan loss provisions, and then plus OBS credit commitment. Interest-bearing assets include loans, due from the central bank, due from banks and other financial institutions, accounts receivable investment, buying back the sale of financial assets, lending to banks and other financial institutions and bonds. OBS credit commitment is classified into guarantee business and credit business. Guarantees are regarded as direct credit substitutes (BCBS 1986), and credit business (e.g. loan commitments) is the most important OBS credit instrument (Chateau 2009). Risk P&L items that are related to credit risk exposure is net interest income and loan impairment loss. The reason is that changes in interest-bearing assets will lead to changes in net interest income, and loan impairment loss should be recorded if there is any indication that loans have suffered an impairment loss (Kwak et al. 2009). Therefore, credit risk P&L is equal to net interest income minus loan impairment loss.

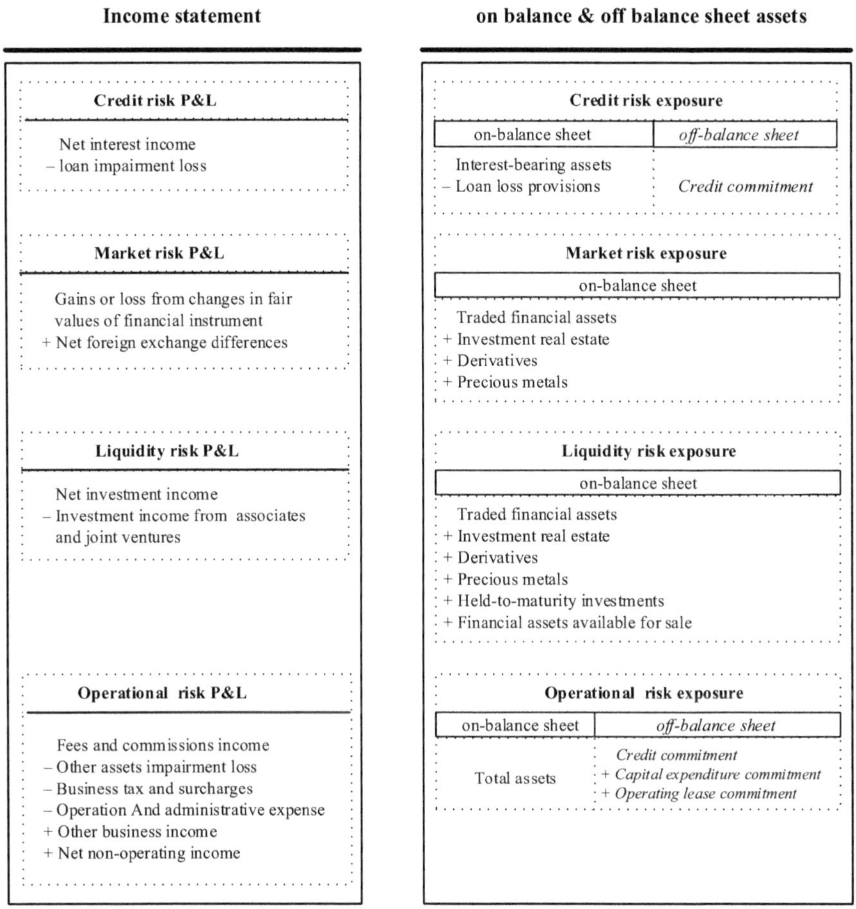

Fig. 8.1 The correspondence between risk types and financial statements

With respect to market risk exposure, it includes traded financial assets, investment real estate, derivatives and precious metals because all these assets are influenced by market factors (i.e. price, interest rate, foreign exchange). Risk P&L items related to market risk exposure are gains or losses from fair values of financial instruments and net foreign exchange differences. The reason is that gains or losses from fair values of financial instruments are affected by fluctuations in the prices of financial instruments. Plus, changes in foreign exchange determine net foreign exchange differences. Thus, the sum of these two accounts is a proxy for market risk P&L.

Liquidity risk exposure includes traded financial assets, investment real estate, derivatives, precious metals, held-to-maturity investments and financial assets available for sale. Net investment income reflects the gains or losses from trading ready to liquidate financial assets. Investment income from associates and joint ventures is generated by long-term equity investments, which is made to control or influence other companies, not to get short-term investment income. Thus, liquidity risk P&L equals net investment income minus investment income from associates and joint ventures.

For operational risk, Rosenberg and Schuermann (2006) deem that all assets and activities of a bank are in some way subject to operational risk. We follow this standpoint that operational risk exposure consists of total on-balance and off-balance sheet assets. The remaining P&L items in the income statement serve as a proxy for operational risk P&L because the operational risk is a typical non-financial risk and represents the volatility of residual earnings, which cannot be categorized into market, credit or liquidity risk.

8.2.2 Procedure of Risk Measurement and Aggregation

Risk P&L items from income statements are not comparable among different banks because banks differ in scale, capital allocation, investment strategy, and management level (Rosenberg and Schuermann 2006). To allow direct comparison across banks, risk P&L needs to be converted into a "risk return" based measure. In accordance with Kretzschmar et al. (2010), we use the data preprocessing method to obtain a specific bank's risk return. The procedure of data preprocessing can be divided into the following three steps:

Firstly, we convert risk P&L into a "risk return" based measure. Since risk P&L is generated by assets that are exposed to risk, an obvious approach for doing this would be to divide risk P&L by assets to yield a return on assets measure. In this chapter, bank assets are defined as risk exposures. Thus, the risk return is the ratio of risk P&L to risk exposure. We then define the marginal risk return for the ith bank, jth risk in period t as

$$r_{i,j,t} = \left(\frac{R_{i,j,t}}{RE_{i,j,t}} \right) \tag{8.1}$$

where $r_{i,j,t}$, $R_{i,j,t}$ and $RE_{i,j,t}$ stand for the risk return, risk P&L and risk exposure of bank i, risk j in period t, respectively.

In the second step, we compute the expected risk return and deviation from risk return. The risk return can be divided into two parts: the expected risk return and deviation from risk return. The expected risk return for a bank is the average risk return over the sample period, which reflects the bank's own characteristics in terms of scale, capital allocation, investment strategy and management level. The deviation from risk return is computed by subtracting the average risk return over the sample period (expected risk return) for each bank, which reflects the macroeconomic background and operating conditions of the whole banking industry. Thus, a bank's risk return is determined by both market and individual information. Specifically, the expected risk return is defined as

$$\overline{r_{i,j}} = \left(\frac{1}{T_i} \sum_{t=1}^{T_i} r_{i,j,t} \right) \tag{8.2}$$

and deviation from risk return as

$$\Delta_{i,j,t} = r_{i,j,t} - \overline{r_{i,j}} \tag{8.3}$$

where bank i is observed for T_i periods. $\overline{r_{ij}}$ denotes the expected risk return for bank i and risk j over the sample T_i period. $\Delta_{i,j,t}$ denotes the deviation from risk return of bank i, risk j in period t.

Finally, we obtain a typical bank's risk returns to model marginal risk distributions. For a typical bank, its risk return is determined by market information and individual information. The market information is composed of all sample banks' deviation from risk return. Thus, by combining all sample banks' deviation from risk return and the typical bank's ($i = k$) expected risk return, we finally compute a typical bank's risk return. Specifically, the typical bank's risk return is written as

$$r_{k,j,t} = \overline{r_{k,j}} + \Delta_{j,t} = \overline{r}_{k,j} + \sum_{i} \Delta_{i,j,t} \tag{8.4}$$

where $r_{k,j,t}$ is the risk return of bank k, risk j in period t. $\overline{r}_{k,j}$ stands for the expected risk return of bank k and risk j. $\Delta_{j,t}$ denotes the summation of deviation from risk return of risk j in period t of all sample banks, which reflects the market information of risk j in period t.

Value-at-Risk (VaR), which has become a standard model for measuring and assessing risk is used to measure marginal risk in this chapter (Huang 2013; Hsu et al. 2012). VaR is defined as a quantile of the distribution of risk returns. Thus, the larger negative value or smaller positive value of VaR corresponds to the higher level of risk (Rosenberg and Schuermann 2006). To aggregate single VaRs into total risk, we adopt the simple summation approach, which is one of the most basic and widely used risk aggregation approaches. Some risk aggregation approaches have emerged so far. Simple summation, var-covar and copula approaches are the three main risk aggregation approaches.

All of them have strengths and weaknesses (Li et al. 2015). The simple summation approach is one of the most basic and widely used approaches to aggregate risk (Rosenberg and Schuermann 2006; Inanoglu and Jacobs 2009; Kretzschmar et al. 2010). It has several features. One is that it is the briefest one that calculates total risk by just adding stand-alone risks. Another is that it is found to be more conservative than other risk aggregation approaches (Embrechts et al. 1999). Such an approach implicitly assumes that all risks are perfectly correlated. That is to say, great losses occur simultaneously, which imposes an upper bound on the true total risk (Dimakos and Aas 2004). Thus, many papers use the simple summation approach to aggregate marginal risks, such as Rosenberg and Schuermann (2006), Inanoglu and Jacobs

(2009) and Kretzschmar et al. (2010). Given that the purpose of this chapter is to analyze the impact of OBS activities on total risk rather than risk aggregation approaches, we adopt the widely used simple summation approach to aggregate different risk types.

Besides, in the use of the simple summation approach to adding marginal risks, the marginal risk weight that represents the marginal risk contribution to total risk should also be considered. Rosenberg and Schuermann (2006) took marginal risk weights into account using the simple summation approach. The total risk, which is referred to as Add-VaR, is the simple weighted summation of marginal risks. The risk weight is the ratio of marginal risk exposure to the total risk exposure (the sum of all marginal risk exposures). Thus, we also use Add-VaR to measure total risk in accordance with Rosenberg and Schuermann (2006). The specific formula of Add-VaR is written as

$$Add - \mathrm{VaR}_{i,t}(\alpha) = \sum_{j} w_{i,j,t} * \mathrm{VaR}_{i,j}(\alpha) \tag{8.5}$$

and the marginal risk weight as

$$w_{i,j,t} = \left(\frac{RE_{i,j,t}}{\sum_j RE_{i,j,t}} \right) = \left(\frac{RE_{i,j,t}}{TRE_{i,t}} \right) \tag{8.6}$$

where $Add - \mathrm{VaR}_{i,t}(\alpha)$ is the total risk in terms of return as a percent of total risk exposure for the ith bank in period t with the $(1 - \alpha)$ confidence level. $\mathrm{VaR}_{i,j}(\alpha)$ is the marginal risk of bank i and risk j under the $(1 - \alpha)$ confidence level. $w_{i,j,t}$ is the marginal risk weight of bank i, risk j in period t. $TRE_{i,t}$ is the sum of different marginal risk exposures and denotes the total risk exposure.

After getting Add-VaR, which represents the loss of unit total risk exposure, we can calculate the total loss by multiplying total risk exposure and Add-VaR. The total loss represents the total risk in terms of losses. It can be written as:

$$TR_{i,t}(\alpha) = Add - VaR_{i,t}(\alpha) * TRE_{i,t} \tag{8.7}$$

where $TR_{r,t}(\alpha)$ represents the total risk in terms of losses for the ith bank in period t with the $(1 - \alpha)$ confidence level.

8.3 Data Description

As discussed in previous chapters, to analyze Chinese banking sector, we collected quarterly panel data over 2007–2018 from early 16 A-share listed Chinese commercial banks (Table 4.1) to ensure the consistency of accounts. Getting rid of exceptional cases (ABC and CEB were listed in 2010. 2007-Q2 data of BOBJ, 2007-Q1 data of BONJ, BONB and CCB are missing.), we finally obtained 743 pieces of valid data to model individual risk distributions.

In the empirical analysis, we use quarterly data to aggregate bank risks. However, OBS items, loan impairment loss and loan loss provision are disclosed only in annual and semi-annual financial reports. Hence, we need to make simple assumptions to obtain quarterly data of these accounts. Specifically, Q1 OBS items are equal to semi-annual OBS items, and Q3 OBS items are equal to annual OBS items. As for loan impairment loss, which is part of assets impairment loss, can be calculated based on known quarterly assets impairment loss. Specifically, we first calculate R, which is a ratio of loan impairment loss to assets impairment loss based on annual and semi-annual data. Then, we calculate the mean value of this ratio over the sample period ($\overline{R}$). Herein, we make a simple assumption that the quarterly R is equal to $\overline{R}$uation>. Thus, the quarterly loan impairment loss is obtained by multiplying $\overline{R}$and quarterly assets impairment loss. Likewise, the quarterly loan loss provision, which is determined by the quality of loans, can be obtained based on the known quarterly loans. Specifically, we define R'as the ratio of loan loss provision to loans and $\overline{R}'$as the mean value of R'over the sample period. Thus, the quarterly loan loss provision is obtained by multiplying$\overline{R}'$with quarterly loans based on the assumption that the quarterly $R'$is equal to $\overline{R}'$.

Among the sample of all 16 early-listed Chinese commercial banks from 2007 to 2018, the amount of financial statements data for a single bank is up to 48, which is too small to perform the empirical analysis. Thus, in order to address the problem of data shortage and provide empirical insights into the total risk of Chinese commercial banks, we construct hypothetical banks following Rosenberg and Schuermann (2006), Kretzschmar et al. (2010) and Alessandri and Drehmann (2010). In particular, we use median assets to characterize hypothetical banks and then a large hypothetical bank and a small hypothetical bank are constructed for comparison. We describe the construction of hypothetical banks in Sect. 7.3 of Chap. 7 in detail. Specifically, the asset size of the large hypothetical bank is the average of the rest three state-owned banks (ICBC, BOC and CCB). The asset size of the small hypothetical bank is the average of the three smallest banks by assets (BOBJ, BONB and BONJ). Using this median approach to constructing hypothetical banks, at the end of 2018, the bank sizes in terms of risk-weighted assets for the large and small hypothetical banks are 14,564 and 1161 billion CNY, respectively.

For either of these two hypothetical banks, the amount of data is 743, which is much larger than that of a real-world bank. Furthermore, the hypothetical banks constructed by us capture the characteristics of real-world banks' asset sizes, so they are the typical banks in Chinese banking system. In a word, performing empirical analysis based on typical hypothetical banks not only addresses the problem of data shortage but also achieves general conclusions.

8.4 Empirical Results

8.4.1 Total Risk Results of Chinese Commercial Banks

The marginal risk distribution is decided by the deviation from risk return and the expected risk return by referring to Eq. (8.4). The deviation from risk return, which reflects the banking industry's macroeconomic background and operating conditions, decides the shape of marginal risk distribution. The expected risk return that reflects a bank's features decides the horizontal axis coordinates of the marginal risk distribution. The characteristics of deviation from risk return are presented numerically in Table 8.1, and the shapes of marginal risk distributions are visually shown in Fig. 8.2.

In terms of return on risk exposure, market risk has the highest volatility (14.48%). Liquidity risk has the fattest tails (kurtosis = 173.65). The volatility (2.48%) of

Table 8.1 Descriptive statistics for deviations from risk return of four marginal risks

	Credit risk	Market risk	Liquidity risk	Operational risk
σ (%)	0.96	14.48	2.48	0.50
Skewness	0.34	2.98	11.29	−0.83
Kurtosis	−0.68	46.26	173.65	1.33

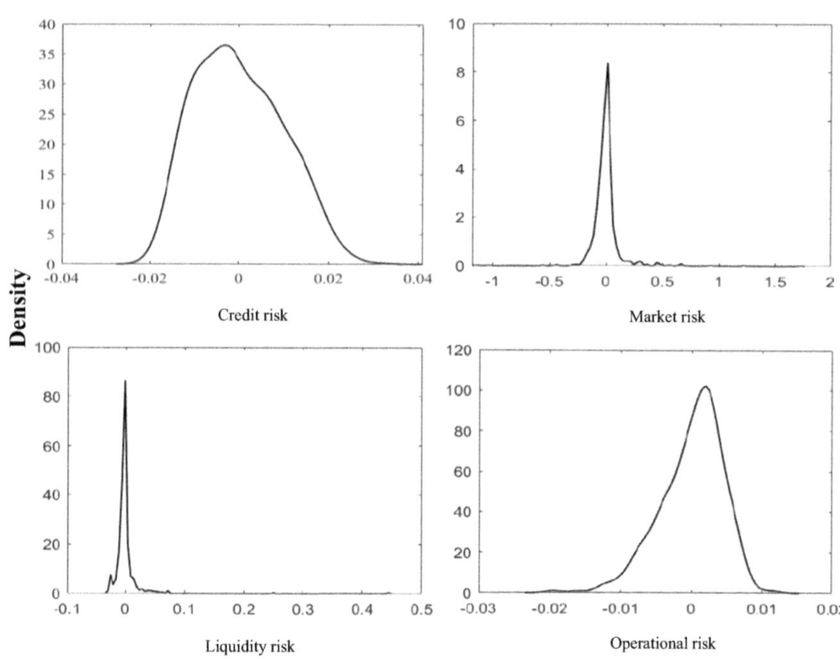

Fig. 8.2 Distributions of deviations from risk return of four marginal risks

liquidity risk and kurtosis of market risk (46.26) come in second. The negative kurtosis present in credit risk (−0.68) and the relatively lower positive kurtosis for operational risk (1.33) suggest that credit risk and operational risk have thinner tails, with volatilities of credit and operational risks being 0.50% and 0.96%, respectively. The shape of operational risk distribution is moderately left-skewed at −0.83, while the other three risk types are right-skewed. Specifically, credit risk is moderately right-skewed at 0.34, market risk is right-skewed at 2.98, and liquidity risk is more significantly right-skewed at 11.29.

Table 8.2 reports two hypothetical banks' expected risk returns. For the large commercial bank, credit, market, liquidity and operational risk expected returns are 0.44%, −9.62%, −0.56%, −0.66%, respectively. The small bank's expected returns for credit (0.40%), liquidity (−0.66%) and operational (−0.76%) risks are less than those of the large bank, while the expected return of market risk (−8.27%) is bigger than that of the large bank.

Since the marginal risk distribution is determined by the expected risk return and deviation from risk return, we finally get the two hypothetical banks' marginal risk distributions. The characteristics of marginal risk distributions are as illustrated in Table 8.3.

Table 8.2 Expected risk returns of the two hypothetical banks (%)

	Large hypothetical bank				Small hypothetical bank			
	Credit risk	Market risk	Liquidity risk	Operational risk	Credit risk	Market risk	Liquidity risk	Operational risk
Expected risk returns	0.44	−9.62	−0.56	−0.66	0.40	−8.27	−0.66	−0.76

Table 8.3 Summary statistics of four marginal risk distributions (%)

	Large hypothetical bank				Small hypothetical bank			
	Credit risk	Market risk	Liquidity risk	Operational risk	Credit risk	Market risk	Liquidity risk	Operational risk
Mean	1.20	−0.77	0.20	−0.21	1.15	0.59	0.09	−0.31
Median	1.20	−0.88	0.16	−0.21	1.16	0.47	0.05	−0.32
0.1th percentile	0.03	−24.70	−2.76	−1.86	0.13	−12.11	−0.16	−1.43
1st percentile	0.21	−19.99	−0.66	−1.55	0.27	−11.28	−0.15	−1.25
2nd percentile	0.21	−19.17	−0.65	−1.43	0.32	−8.40	−0.15	−1.08
5th percentile	0.34	−13.28	−0.39	−0.91	0.49	−6.31	−0.01	−1.04

It is clear that the statistics of marginal risk distributions are of difference between the large and small banks. In particular, at 99.9% confidence level, mean values of the large bank's credit, market, liquidity and operational risks are 1.20%, −0.77%, 0.20% and −0.21%, respectively. For the small bank, mean values of credit, market, liquidity and operational risks are 1.15%, 0.59%, 0.90% and −0.31%, respectively. As for median values, the large bank's credit risk has the largest median (1.20%) while the median of operational risk (−0.21%) is between that of liquidity risk (0.16%) and market risk (−0.88%). Regarding the small bank, median of credit risk (1.16%) is still the largest while the median of market risk (0.47%) ranks latter. The smallest median is −0.32% of the operational risk, and the median of liquidity risk is 0.05%.

The last four columns of Table 8.3 are VaR values of individual risks at different confidence levels. Since the larger negative value or smaller positive value of VaR corresponds to the higher level of risk, the empirical results suggest that the large bank's market risk is higher while the other three risk types (credit, liquidity and operational risks) are lower compared with the small bank. Specifically, the large bank's negative value of market risk VaR is larger than that of the small bank, so that the large bank's market risk is higher. As for liquidity risk and operational risk, the large bank's negative values of VaRs are smaller than those of the small bank. The large bank's positive value of credit risk is larger than that of the small bank. Thus, the large bank's credit, liquidity and operational risks are lower. For example, at the 0.1th percentile, the large bank's market risk VaR is −19.99%, whose negative value is larger than that of the small bank (−11.28%). While the large bank's liquidity risk and operational risk VaRs are −2.76% and −1.86%, respectively, whose negative values are smaller than those of the small bank (liquidity risk: −0.16%; operational risk: −1.43%). The large bank's positive value of credit risk VaR (0.03%) is smaller than that of the small bank (0.13%). Therefore, the large bank's market risk is higher while credit, liquidity and operational risks are lower compared with the small bank.

After getting VaR values of marginal risks, we then calculate the total risk by just adding single VaRs. According to Eq. (8.5), Add-VaR, the total risk in terms of return as a percent of total risk exposure, is the weighted simple summation of marginal risks. In 2018, the large bank's marginal risk weights are 44.54%, 0.94%, 7.96% and 46.56% for credit, market, liquidity and operational risks, respectively. The small bank's credit, market, liquidity and operational risk weights are 45.24%, 0.49%, 7.69% and 46.58%, respectively. Then, the total loss, which is the total risk in terms of losses, is calculated based on Eq. (8.7). For simplicity in what follows, we shall refer to both Add-VaR and total loss as the total risk. Our quarterly financial statements data enable a quarterly view of total risk while the typical horizon of losses is one year. To transform the quarterly total loss into the annual total loss, we apply the *square-root-of-time rule,* which is commonly used to scale an estimated quantile of a return distribution to a lower frequency T by the multiplication of $\sqrt{T}$ (Danielsson and Zigrand 2006). Table 8.4 gives a summary overview of the two hypothetical banks' total risks in 2018.

As shown in Table 8.4, there is no significant difference between the Add-VaR values of these two banks, while the total loss of the large commercial bank is bigger because of its larger scale. Specifically, at 99.9% confidence level, Add-VaR values

Table 8.4 2018 Add-VaRs and total losses of the two hypothetical banks

Confidence level (%)	Large hypothetical bank				Small hypothetical bank			
	99.9	99	98	95	99.9	99	98	95
Total risk								
(Add-VaR) (%)	−1.61	−1.21	−1.12	−0.65	−0.93	−0.77	−0.59	−0.43
Total risk								
(Annual losses in billion CNY)	983.90	738.94	686.68	399.76	42.07	34.74	26.39	19.20

of the large and small banks are −1.61% and −0.93%, respectively. The large bank may suffer an annual total loss of 984 billion CNY while the small bank's annual total loss is equal to 42 billion CNY.

8.4.2 The Impact of OBS Activities on Total Risk

In this section, we empirically test whether the Chinese commercial bank's total risk is affected by OBS activities. Compared with the existing financial statements based risk aggregation framework, which only aggregates on-balance risks, our improved framework can capture both on-balance and off-balance sheet risks. Therefore, by comparing risk aggregation results estimated by these two frameworks, we first examine the overall impact of OBS activities on the bank's total risk and then further study the individual impact of each OBS risk type on total risk. To our best knowledge, no existing studies examine the effect of OBS activities on the Chinese commercial bank's total risk.

Although there are some studies on the correlation between banks' risks and OBS activities, there is no consensus thus far. Traditionally, OBS activities have been seen as a risk-reducing tool (Hassan et al. 1994). In contrast to findings that OBS items are negatively correlated with banks' risks, some believe that banking institutions heavily involved in OBS activities are characterized by higher risks (Calmès and Théoret 2010; Papanikolaou and Wolff 2014).

By comparing the total risk with and without OBS items in 2018 (Fig. 8.3), we empirically prove that the risk entailed in OBS activities affects the total risk of Chinese commercial banks and the impact depends on bank size. In particular, at 99.9% confidence level, the large bank's total risk decreases from 984 billion CNY to 814 billion CNY while the small bank's total risk decreases from 42 billion CNY to 40 billion CNY after taking OBS items into risk aggregation. Therefore, the entire OBS activities exert a negative effect on both the large bank's and small bank's total risk, especially for the large bank. Ignoring OBS items in risk aggregation will overestimate the large bank's and small bank's total risk. With the development of off-balance sheet business, the effect of reducing the total risk of large banks is more significant.

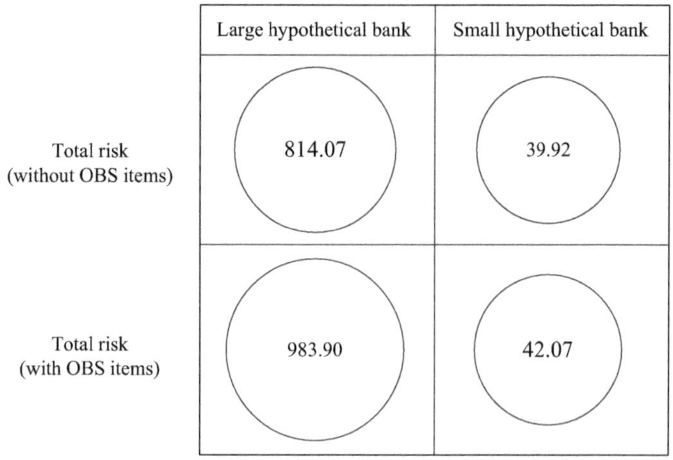

Fig. 8.3 Total risk with and without OBS items in 2018 (Unit: billion CNY)

The reason why the overall impact of OBS activities on the bank's total risk depends on bank size may be that the ability of OBS risk management is different between banks. OBS items have both risk-reducing as well as risk-increasing attributes. The ability of OBS risk management determines the net impact of OBS items on bank risks (Khasawneh et al. 2012). Compared with the large bank, the small bank engages in more risky OBS activities without experience and expertise (Mercieca et al. 2007). Thus, OBS items increase the small bank's total risk while decreasing the large bank's total risk. In particular, stated-owned banks play a leading role in the traditional deposits and loans market. Zhao and Jian (2013) have also confirmed that the larger the bank, the stronger the ability of bank profitability. Since a keener competition leads to greater risk-taking behaviors (Hellmann et al. 2000), the small bank engages in more risky OBS activities for pursuing higher profit. For example, the use of OBS derivatives as speculation rather than hedging tools increases the riskiness of the small bank. Furthermore, limited knowledge on markets and OBS transactions hampers the small bank's performance. Therefore, OBS items are negatively linked to the Chinese large bank's total risk while positively linked to the small bank's total risk.

However, the findings in Fig. 8.3 also reveal that the gap between total risk with and without OBS items is not obvious, which weakens the need of incorporating OBS items into risk aggregation framework. Thus, we then explain why the difference caused by OBS items in risk aggregation is not apparent. To show the change of total risk brought about by OBS activities, we compute a change rate that is the proportional change of total risk ($\Delta TR\%$) using total risk with OBS activities (TR_{OBS}) versus total risk without OBS activities (TR) : $(TR_{OBS} - TR)/TR$.

According to (Eq. 8.7), total risk is the product of Add-VaR and total risk exposure. Consequently, the specific equation for the change rate of total risk after considering OBS items can be written as

$$
\begin{aligned}
\Delta TR\% &= \Delta TR/TR = (TR_{OBS} - TR)/TR \\
&= \left[\mathrm{TRE}_{\mathrm{OSS}} * (Add - VaR_{\mathrm{CBS}}) - TRE^{*}(Add - VaR)\right]/TR \\
&= \left\{TRE^{*}(1 + \Delta TRE\%)^{*}(Add - VaR)[1 + \Delta(Add - VaR)\%]\right. \\
&\quad \left. - TRE^{*}(Add - VaR)\right\}/TR \\
&= \left\{TRE^{*}(Add - VaR)^{*}[(1 + \Delta TRE\%) * (1 + \Delta(Add - VaR)\%) - 1]\right\}/TR \\
&= TR^{*}[(1 + \Delta TRE\%) * (1 + \Delta(Add - VaR)\%) - 1]/TR \\
&= (1 + \Delta TRE\%) * (1 + \Delta(Add - VaR)\%) - 1 \qquad (8.8)
\end{aligned}
$$

where $\Delta TR\%$, $\Delta TRE\%$, and $\Delta(Add - VaR)\%$ are the change rate of total risk, total risk exposure and Add-VaR, respectively. After incorporating OBS activities into risk aggregation, the total risk exposure will increase while Add-VaR will decrease. Therefore, $\Delta TRE\%$ is the positive value while $\Delta(Add - VaR)\%$ is the negative value. With larger absolute values of $\Delta TRE\%$ and $\Delta(Add - VaR)\%$, the absolute value of $\Delta TR\%$ will become larger. For example, $\Delta TR\%$ is equal to -1% if $\Delta TRE\%$ is 10% and $\Delta(Add - VaR)\%$ is -10%. However, if $\Delta TRE\%$ increases to 40% and $\Delta(Add - VaR)\%$ decreases to -40%, $\Delta TR\%$ is equal to -16%.

Figure 8.4 visually shows the change rates of total risk exposure, Add-VaR and total risk after incorporating OBS activities into risk aggregation. Specifically, the large bank's total risk declines by 20.86% and the small bank's total risk decreases by 5.45%. According to Eq. (8.8), the unapparent change rate of total risk is caused by smaller absolute values of $\Delta TRE\%$ and $\Delta(Add - VaR)\%$. There are two reasons why the absolute values of $\Delta TRE\%$ and $\Delta(Add - VaR)\%$ are smaller. One is that only a part of OBS items is taken into risk measurement. Another is that the current scale of Chinese banks' OBS activities is still small compared with western countries' banks.

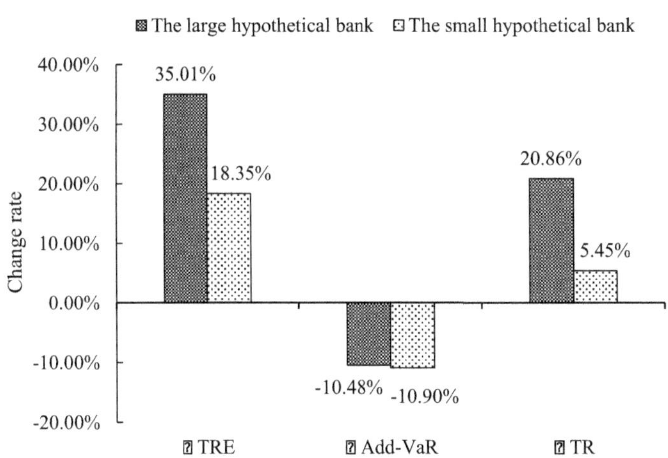

Fig. 8.4 The change rates caused by incorporating OBS items into risk aggregation

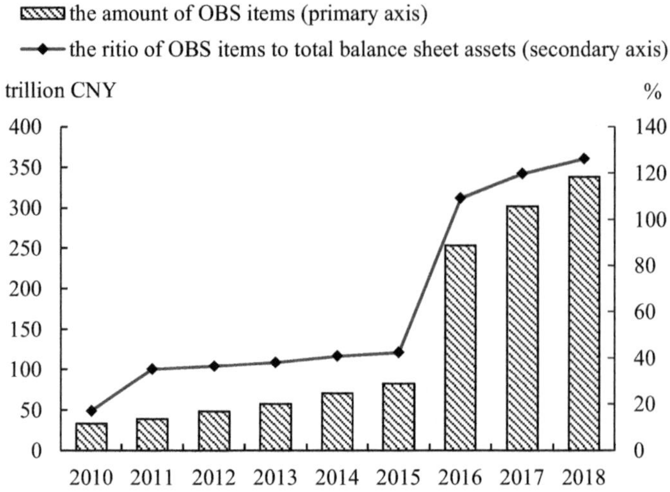

Fig. 8.5 The growth of Chinese banking OBS activities

Based on available data, the OBS items we consider in risk aggregation account for 35.01% and 18.35% of total on-balance sheet assets for the large bank and small bank, respectively. However, according to the *China Financial Stability Report 2019*, at the end of 2018, OBS items accounted for 125% of total on-balance sheet assets. Therefore, the absolute values of $\Delta TRE\%$ and $\Delta(Add - VaR)\%$ will get larger after considering all OBS items.

In addition, the continuous growth of OBS activities in Chinese banking system will enlarge the absolute values of $\Delta TRE\%$ and $\Delta(Add - VaR)\%$. According to the *China Financial Stability Report 2019,* OBS items of Chinese banking system continued to rise over 2010–2018 (Fig. 8.5). Specifically, OBS activities of Chinese banks increased from 33 trillion CNY to 354 trillion CNY, and the percentage of OBS items accounted for total balance sheet assets increased from 35 to 125%. In the past decade, off-balance sheet business has expanded rapidly in Chinese banking industry. Thus, ignoring off-balance sheet items in bank risk aggregation will lead to greater deviation of risk aggregation results.

All in all, the increasingly standardized disclosure requirements for OBS items and the continuous growth of OBS activities in Chinese banking system will enlarge the changes of total risk exposure and Add-VaR after taking OBS items into risk aggregation, which further result in a more apparent difference between the total risk with and without OBS items. Thus, incorporating OBS items into risk aggregation framework is reasonable and necessary.

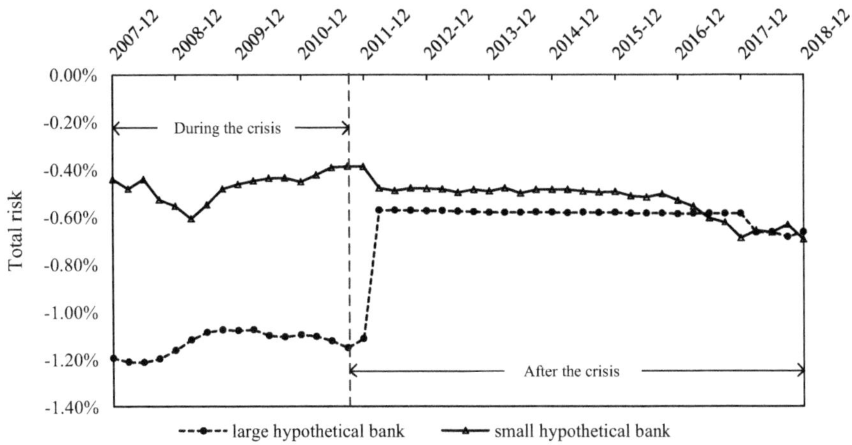

Fig. 8.6 The changes of total risks of Chinese commercial banks during and after the subprime crisis. The larger negative value corresponds to the higher level of total risk

8.4.3 Total Risk Transformation in the Subprime Crisis

After obtaining the total risks of Chinese commercial banks, in the following text, we further study how Chinese banks' risks are affected by the subprime crisis. Indeed, some studies have found that the financial crisis affected the bank risk-taking behavior, and bank risks were related to the phase of the business cycle (Delis and Kouretas 2011; Shim 2013). Therefore, in this chapter, we study how the subprime crisis affected Chinese commercial banks' risks by dividing the entire sample into two subsets, during and after the subprime crisis. In accordance with Zhu et al. (2015), the subprime crisis began at 2007-Q4 and ended at 2011-Q3.

As illustrated in Fig. 8.6, the total risk (Add-VaR at 99.9% confidence level) is significantly larger during times of crisis for both large and small banks. Just like most banks across the world which had plunged into severe risk (Dias and Ramos 2014), Chinese commercial banks had higher risks during the subprime crisis.

To further study why total risk became larger when the subprime crisis broke out, we analyze risk weights and marginal risks because the total risk is decided by both of them (following (Eq. 8.5)). Figure 8.7 visually shows that the marginal risk weights almost remain unchanged. In other words, the outbreak of the subprime crisis did not significantly affect the business mix of Chinese commercial banks. Thus, the change of marginal risk weights is not the main cause for the larger total risk during the crisis. Then, we show the changes in four marginal risks of Chinese commercial banks during and after the crisis in Fig. 8.8. As noted above, the larger negative value or smaller positive value corresponds to higher marginal risk. It is clear that for both large and small banks, market and liquidity risks experienced a sharp increase while changes in credit and operational risks were not obvious with the outbreak of the

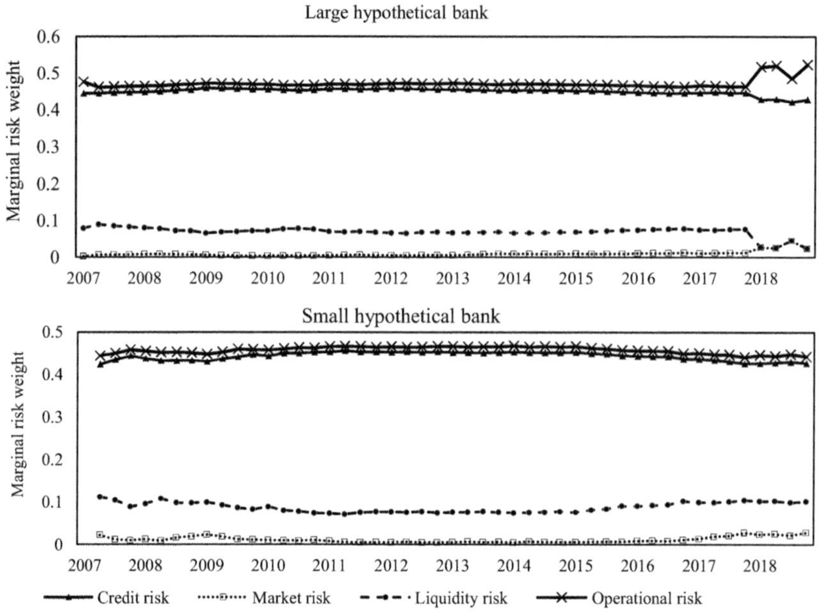

Fig. 8.7 The four marginal risk weights during and after the crisis

subprime crisis. The VaR values of marginal risks reported in Table 8.5 also support this finding precisely.

Table 8.5 reports VaR values of marginal risks at a 99.9% confidence level during and after the subprime crisis. For both large and small banks, there are slight changes in VaR values of credit and operational risks while significant changes in VaR values of market and liquidity risks. In particular, when the subprime crisis broke out, the large bank's credit risk VaR slightly decreased from 0.37 to 0.32% and operational risk VaR experienced a modest reduction from −0.69 to −0.88%. However, it is important to note that market risk VaR drastically dropped from −9.00 to −24.39%, and liquidity risk VaR substantially dropped from −0.54 to −1.63%. As for the small bank, credit risk and operational risk changed slightly, with VaR values of credit risk and operational risk being 0.33% and −0.98% during the crisis and 0.26% and −0.82% after the crisis, respectively. However, the small bank's market risk VaR substantially dropped from −9.99 to −21.55% and liquidity risk VaR decreased from −0.60 to −1.81% with the outbreak of the subprime crisis.

Therefore, we can conclude that the significant increase of liquidity risk and market risk is the main cause for the larger total risk during the subprime crisis. Furthermore, our empirical results are consistent with findings in Alexander et al. (2013) and Cornett et al. (2011). Alexander et al. (2013) found that banks across the world suffered vast trading losses during the subprime crisis. Cornett et al. (2011)

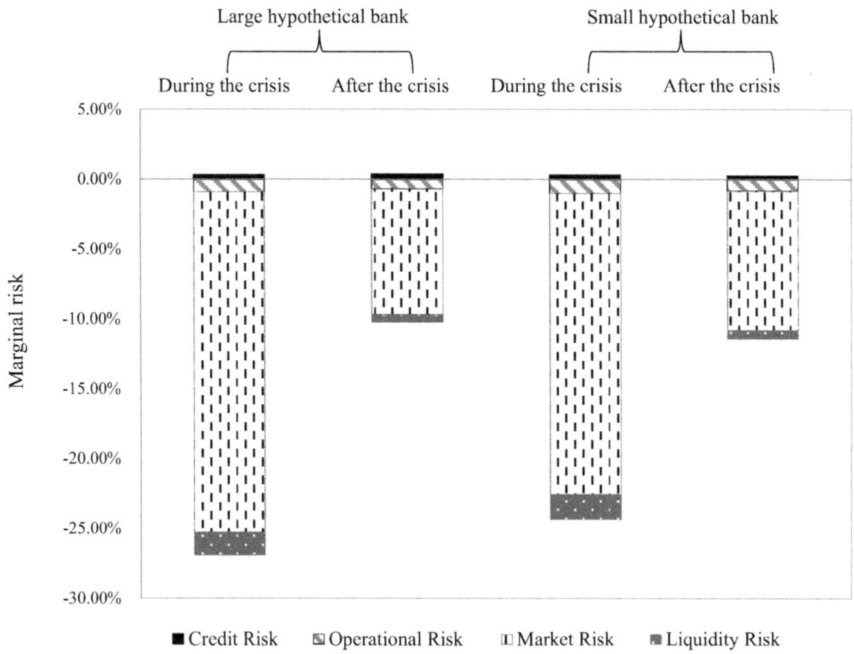

Fig. 8.8 The changes in four marginal risks of Chinese commercial banks during and after the subprime crisis. The larger negative value corresponds to the higher level of marginal risk

Table. 8.5 The comparison of marginal risks during and after the subprime crisis

Marginal risk (99.9% VaR)	Large hypothetical bank		Small hypothetical bank	
	During the crisis	After the crisis	During the crisis	After the crisis
Credit risk (%)	0.32	0.37	0.33	0.26
Operational risk (%)	−0.88	−0.69	−0.98	−0.82
Market risk (%)	−24.39	−9.00	−21.55	−9.99
Liquidity risk (%)	−1.63	−0.54	−1.81	−0.60

concluded that liquidity dried up during the bad years of 2007–2009 and liquidity risk was a challenge to a bank in times of stress.

8.5 Conclusions

In this chapter, we improve the financial statements based risk aggregation framework by incorporating OBS activities into risk aggregation, which allows us to capture both on-balance and off-balance sheet risks simultaneously. In the empirical analysis, we apply this improved framework to aggregate credit, market, liquidity and operational risks by using a sample of all 16 early-listed Chinese listed commercial banks for 2007–2018. Then we empirically study whether the overall impact of OBS activities and the individual impact of each OBS risk type on total risk depend on bank size by constructing two typical Chinese commercial banks. Moreover, this research divides the samples into two subsets to find out the transformation of Chinese banks' risks during and after the subprime crisis.

Our empirical results show that the total risk of Chinese commercial banks is affected by OBS activities. Specifically, the entire OBS activities are negatively related with both the large and small banks' total risk. Furthermore, the risk reduction of large commercial banks due to off-balance sheet business is more significant. Hence, since the scale of off-balance sheet business of large commercial banks accounts for a larger proportion of the on-balance sheet assets, the large bank's total risk is more overestimated than the small bank's total risk if ignoring off-balance sheet items. Besides, the risk transformation analysis for Chinese commercial banks suggests that it is the increase of liquidity risk and market risk that leads to the larger total risks for both large and small banks during the subprime crisis.

However, this chapter has several limitations. First, the difference between total risk with and without OBS items is not obvious. However, the development of OBS activities and the increasingly standardized disclosure requirements for OBS items will enlarge the deviation if OBS items are ignored in risk aggregation. Second, the correspondence between risk types and financial statements is kind of rough. For example, net interest income bears credit risk and market risk simultaneously. Moreover, whether all assets are subject to operational risk is still open to question. In future studies, the employment of other information may help calibrate the corresponding relationship to some extent.

References

Acerbi C, Scandolo G (2008) Liquidity risk theory and coherent measures of risk. Quant Finance 8(7):681–692. https://doi.org/10.1080/14697680802373975

Acharya VV, Richardson M (2009) Causes of the financial crisis. Crit Rev 21(2–3):195–210. https://doi.org/10.1080/08913810902952903

Alessandri P, Drehmann M (2010) An economic capital model integrating credit and interest rate risk in the banking book. J Bank Finance 34(4):730–742. https://doi.org/10.1016/j.jbankfin.2009.06.012

Alexander GJ, Baptista AM, Yan S (2013) A comparison of the original and revised Basel market risk frameworks for regulating bank capital. J Econ Behav Organ 85:249–268. https://doi.org/10.1016/j.jebo.2012.04.007

Bae KH, Tan H, Welker M (2008) International GAAP differences: the impact on foreign analysts. Account Rev 83(3):593–628. https://doi.org/10.2308/accr.2008.83.3.593

Basel Committee on Banking Supervision (1986) The management of banks' off-balance sheet activities exposures. Bank for International Settlements, Basel

Basel Committee on Banking Supervision (1988) International convergence of capital measurement and capital standards. Bank for International Settlements, Basel

Basel Committee on Banking Supervision (1996) Amendment to the capital accord to incorporate market risks. Bank for International Settlements, Basel

Basel Committee on Banking Supervision (2003) Trends in risk integration and aggregation. Bank for International Settlements, Basel
Basel Committee on Banking Supervision (2006) International convergence of capital measurement and capital standards: a revised framework. Bank for International Settlements, Basel
Basel Committee on Banking Supervision (2008) Principles for sound liquidity risk management and supervision. Bank for International Settlements, Basel
Basel Committee on Banking Supervision (2010) Basel III: a global regulatory framework for more resilient banks and banking systems. Bank for International Settlements, Basel
Blundell-Wignall A, Atkinson P (2010) Thinking beyond Basel III. OECD J: Finan Market Trends 2010(1):9–33. https://doi.org/10.1787/fmt-2010-5km7k9tpcjmn
Boyd JH, Gertle M (1994) Are banks dead? Or are the reports greatly exaggerated? Q Rev
Brunnermeier MK (2009) Deciphering the liquidity and credit crunch 2007–2008. J Econ Perspect 23(1):77–100. https://doi.org/10.1257/jep.23.1.77
Calmès C, Théoret R (2010) The impact of off-balance-sheet activities on banks returns: an application of the ARCH-M to Canadian data. J Bank Finance 34(7):1719–1728. https://doi.org/10.1016/j.jbankfin.2010.03.017
Chateau JPD (2009) Marking-to-model credit and operational risks of loan commitments: a Basel-2 advanced internal ratings-based approach. Int Rev Financ Anal 18(5):260–270. https://doi.org/10.1016/j.irfa.2009.07.001
Chavez-Demoulin V, Embrechts P, Nešlehová J (2006) Quantitative models for operational risk: extremes, dependence and aggregation. J Bank Finance 30(10):2635–2658. https://doi.org/10.1016/j.jbankfin.2005.11.008
Cornett MM, McNutt JJ, Strahan PE, Tehranian H (2011) Liquidity risk management and credit supply in the financial crisis. J Financ Econ 101(2):297–312. https://doi.org/10.1016/j.jfineco.2011.03.001
Danielsson J, Zigrand JP (2006) On time-scaling of risk and the square-root-of-time rule. J Bank Finance 30(10):2701–2713. https://doi.org/10.1016/j.jbankfin.2005.10.002
Delis MD, Kouretas GP (2011) Interest rates and bank risk-taking. J Bank Finance 35(4):840–855. https://doi.org/10.1016/j.jbankfin.2010.09.032
DeYoung R, Rice T (2004) Noninterest income and financial performance at U.S. commercial banks. Financ Rev 39(1):101–127. https://doi.org/10.1111/j.0732-8516.2004.00069.x
Dias JG, Ramos SB (2014) The aftermath of the subprime crisis: a clustering analysis of world banking sector. Rev Quant Financ Acc 42(2):293–308. https://doi.org/10.1007/s11156-013-0342-3
Dimakos XK, Aas K (2004) Integrated risk modelling. Stat Model 4(4):265-277. https://doi.org/10.1191/1471082X04st079oa
Drehmann M, Sorensen S, Stringa M (2010) The integrated impact of credit and interest rate risk on banks: a dynamic framework and stress testing application. J Bank Finance 34(4):713–729. https://doi.org/10.1016/j.jbankfin.2009.06.009
Embrechts P, McNeil AJ, Straumann D (1999) Correlation: pitfalls and alternatives. Risk 12:69–71
Hassan MK, Karels GV, Peterson MO (1994) Deposit insurance, market discipline and off-balance sheet banking risk of large U.S. commercial banks. J Bank Finance 18(3):575–593. https://doi.org/10.1016/0378-4266(94)90010-8
Hellmann TF, Murdock KC, Stiglitz JE (2000) Liberalization, moral hazard, in banking and prudential regulation: are capital requirements enough? Am Econ Rev 90:147–165. https://doi.org/10.1257/aer.90.1.147
Hou X, Wang Q, Li C (2015) Role of off-balance sheet operations on bank scale economies: evidence from China's banking sector. Emerg Mark Rev 22:140–153. https://doi.org/10.1016/j.ememar.2014.10.001
Hsu CP, Huang CW, Chiou WJP (2012) Effectiveness of copula-extreme value theory in estimating value-at-risk: empirical evidence from Asian emerging markets. Rev Quant Financ Acc 39(4):447–468. https://doi.org/10.1007/s11156-012-0283-2

Huang AYH (2013) Value at risk estimation by quantile regression and kernel estimator. Rev Quant Financ Acc 41(2):225–251. https://doi.org/10.1007/s11156-012-0308-x

Inanoglu H, Jacobs M (2009) Models for risk aggregation and sensitivity analysis: an application to bank economic capital. J Risk Financ Manag 2(1):118–189. https://doi.org/10.3390/jrfm2010118

Karim D, Liadze I, Barrell R, Davis EP (2013) Off-balance sheet exposures and banking crises in OECD countries. J Financ Stab 9(4):673–681. https://doi.org/10.1016/j.jfs.2012.07.001

Khasawneh AY, Khrawish HA, Khrisat FA (2012) The determinants of OBS activities in Jordan banking system: panel data analysis. Eur J Econ Finance Adm Sci 47:30–42

Kretzschmar G, McNeil AJ, Kirchner A (2010) Integrated models of capital adequacy—why banks are undercapitalized. J Bank Finance 34(12):2838–2850. https://doi.org/10.1016/j.jbankfin.2010.02.028

Kuritzkes A, Schuermann T (2007) What we know, don't know and can't know about bank risk: a view from the trenches. The known the unknown and the unknowable in financial risk management. Princeton University Press, Princeton

Kwak W, Lee HY, Eldridge SW (2009) Earnings management by Japanese bank managers using discretionary loan loss provisions. Rev Pac Basin Financ Mark Policies 12(01):1–26. https://doi.org/10.1142/S0219091509001526

Li J, Yi S, Zhu X, Feng J (2012) Mutual information based copulas to aggregate banking risks. In: 2012 Fifth international conference on business intelligence and financial engineering, Lanzhou

Li J, Zhu X, Chen J, Gao L, Feng J, Wu D, Sun X (2014) Operational risk aggregation across business lines based on frequency dependence and loss dependence. Math Prob Eng 2014:404208. https://doi.org/10.1155/2014/404208

Li J, Zhu X, Lee CF, Wu D, Feng J, Shi Y (2015) On the aggregation of credit, market and operational risks. Rev Quant Financ Acc 44(1):161–189. https://doi.org/10.1007/s11156-013-0426-0

Mercieca S, Schaeck K, Wolfe S (2007) Small European banks: benefits from diversification? J Bank Finance 31(7):1975–1998. https://doi.org/10.1016/j.jbankfin.2007.01.004

Mustika G, Suryatinc E, Hall MJB, Simper R (2015) Did Bank Indonesia cause the credit crunch of 2006–2008? Rev Quant Financ Acc 44(2):269–298. https://doi.org/10.1007/s11156-013-0406-4

Papanikolaou NI, Wolff CCP (2014) The role of on-and off-balance-sheet leverage of banks in the late 2000s crisis. J Financ Stab 14:3–22. https://doi.org/10.1016/j.jfs.2013.12.003

Rajan RG (2006) Has finance made the world riskier? Eur Financ Manag 12(4):499–533. https://doi.org/10.1111/j.1468-036X.2006.00330.x

Rosenberg JV, Schuermann TA (2006) General approach to integrated risk management with skewed, fat-tailed risks. J Financ Econ 79(3):569–614. https://doi.org/10.1016/j.jfineco.2005.03.001

Saito Y (2012) The demand for accounting information: young NASDAQ listings versus S&P 500 NYSE listings. Rev Quant Financ Acc 38(2):149–175. https://doi.org/10.1007/s11156-010-0223-y

Seow GS, Tam K (2002) The usefulness of derivative-related accounting disclosures. Rev Quant Finance Account 18(3):273–291. https://doi.org/10.1023/A:1015392302022

Shim J (2013) Bank capital buffer and portfolio risk: the influence of business cycle and revenue diversification. J Bank Finance 37(3):761–772. https://doi.org/10.1016/j.jbankfin.2012.10.002

Zhao S, Jian J (2013) Size does matter for profitability! Testing economies of scale in Chinese banking industry. J Appl Sci 13(17):3510. https://doi.org/10.3923/jas.2013.3510.3515

Zhu X, Xie Y, Li J, Chen J, Yang S, Sun X, Wu D (2015) Change point detection for subprime crisis in American banking: from the perspective of risk dependence. Int Rev Econ Finance 38:18–28. https://doi.org/10.1016/j.iref.2014.12.011

Chapter 9
Analysis of Textual Risk Disclosures in Financial Statements

9.1 Introduction

Economic theory tells us that bank risks are affected by some risk factors (Jarrow and Turnbull 2000). Bank risks can be described as profits and losses due to movements in risk factors. Credit risk, market risk, operational risk and liquidity risk are related to different factors. They are often affected by various specific factors in reality. For example, specifically, credit risk is associated with counterparty failure (Li et al. 2015). Thus, uncertainties about risk profit and loss (P&L) are related to uncertainties about risk factors, which means risk factors can provide a logical answer to the risk P&L (Rosenberg and Schuermann 2006).

A comprehensive selection of risk factors is of the utmost importance for explaining risk P&L (Embrechts et al. 2013). Risk factors can capture bank risks well if they can explain most of the P&L uncertainty. Suppose they explain only a small fraction of the risk P&L uncertainty. In that case, the residuals will have to be examined to see whether many more risk factors exist that are worth investigating and including in the risk analysis (Alexander and Pezier 2003; Breuer et al. 2010). Omitting risk factors may lead to bias in estimating bank risks.

Since bank risk factors are essential for measuring bank risks, many studies have made contributions to the identification of bank risk factors (Grundke 2010). Specifically, Jarrow and Turnbull (2000), Alexander and Peizer (2003), Medova and Smith (2005), Rosenberg and Schuermann (2006), Aas et al (2007), Grundke (2009), Kretzschmar, et al (2010), Breuer et al (2010), Grundke (2010) and Bellini (2013) determined specific macroeconomic risk factors, including interest rate, credit spread, exchange rate, equity market index, gross domestic product and many more. Further, local competition (Chari and Gupta 2008), local culture (Li and Guisinger 1992), the degree of regulatory, monetary, and legal complexity (Berger et al. 2004), mergers and acquisitions (Alibux 2007; Hagendorff and Keasey 2009), the degree of economic and political instability (Brewer and Rivoli 1990), the extent of market imperfections, and asymmetric information problems (Buch and DeLong 2004; Gleason et al. 2006) are also used as risk factors in measuring bank risks.

J. Li et al., *Financial Statements-Based Bank Risk Aggregation*, Innovation in Risk Analysis, https://doi.org/10.1007/978-981-19-0408-0_9

However, there is less consensus on which factors affect bank risks in previous studies. Bank risk factors selected by different studies are various. To our best knowledge, almost no prior work can fully determine how many risk factors affect bank risks, while researchers have simultaneously realized that many potential risk factors are excluded from risk estimation. Since identifying risk factors for measuring bank risks is difficult and complex, it is infeasible to comprehensively and accurately identify bank risk factors depending on the researchers' adjustments or traditional theories in previous studies. Hence, a new way to comprehensively identify bank risk factors is needed.

Recently, managers and researchers have been interested in corporate qualitative textual risk disclosures (Bao and Datta 2014), which consist of massive and comprehensive information on risks that corporates face. Specifically, beginning in 2005, the U.S. Securities and Exchange Commission (SEC) required public companies to add a separate section 1A in their SEC Form 10-K to discuss "the most significant factors that make the offering speculative or risky" (SEC 2005). The qualitative textual risk disclosures reported in financial statements are important sources of corporate risk factors (Campbell et al. 2014). Thus, analyzing textual risk disclosures reported in financial statements of the banking industry is a feasible way to solve the problem of comprehensive identification of bank risk factors that have not yet been solved. Some researchers have provided insight into analyzing corporate textual risk disclosures (Huang and Li 2011; Bao and Datta 2014; Dyer et al. 2017; Miller 2017). However, no prior work focuses on analyzing the textual risk disclosures of the banking industry.

To discover bank risk factors from textual risk disclosures reported in section 1A of Form 10-K, an automatic text analysis approach is needed. Researchers have adopted the supervised, dictionary and unsupervised text classification approaches based on the characteristics of textual risk factors. In particular, Huang and Li (2011) proposed a supervised learning approach called the multi-label categorical K-nearest neighbor (ML-CKNN) to categorize risk factors. Campbell et al (2014) use a predefined dictionary to quantify risk factors disclosed in Form 10-K filings. However, the above two text mining approaches need to predefine a list of risk factors reported in financial disclosures, which might be hard to derive beforehand in most cases (Zhu et al. 2016). Huang and Li (2011) set the number of risk factors at 25 while Campbell et al (2014) predefined five risk factor types based on their subjective judgments, which leads to the problem that some critical risk factor types may be left out.

To address the problem of incomplete identification of risk factors, Bao and Datta (2014) proposed an unsupervised Sentence Latent Dirichlet Allocation (Sent-LDA) to discover risk factors comprehensively. However, the unsupervised algorithm makes researchers stay away from data, leading to biased classification results without appropriate human interpretation and adjustment (Miller 2017). The semi-supervised text mining approach can achieve higher accuracy of classification results by using a large amount of unlabeled data, together with the labeled data (Zhu 2007). However, as far as we know, few previous works attempt to apply semi-supervised approaches to classify textual risk disclosures of Form 10-K.

Therefore, this chapter aims to comprehensively and accurately identify bank risk factors by proposing a new semi-supervised text mining naive collision algorithm. Compared with unsupervised text mining approaches for Form 10-K, the semi-supervised text mining approach proposed by us can give a higher accuracy classification result of bank risk factors. In the experiment, based on 59,418 textual risk disclosures reported in section 1A from 2189 SEC Form 10-K of U.S. commercial banks for 2010–2016, we identify 21 risk factors affecting bank risks. To determine which of these 21 identified bank risk factors are unique to the banking industry, we compare the 21 bank risk factors with risk factors discovered in previous studies. Furthermore, we analyze the importance of each risk factor to find which factors have more significant impacts on bank risks, aiming to make the bank risk measurement more efficient. Last, we made an empirical comparison between our proposed semi-supervised text mining approach and a typical unsupervised text mining approach to show whether our approach can give a higher accuracy classification result.

The remainder of this chapter is organized as follows. Section 9.2 introduces the proposed semi-supervised text mining naive collision algorithm. Section 9.3 describes the data collection. Section 9.4 discusses the discovered bank risk factors. Section 9.5 concludes this chapter.

9.2 The Naive Collision Algorithm

This section introduces our proposed semi-supervised text mining algorithm, called the naive collision algorithm, for classifying textual risk disclosures. We observe that almost all risk factor disclosures consist of a summary heading and detailed explanations. However, applying text mining approaches to both headings and descriptions led to worse performance in prior studies (Huang and Li 2011). Thus, our proposed algorithm is designed for classifying headings of textual risk disclosures. Furthermore, following Bao and Datta (2014), we assume that all words in a heading are sampled from the same topic.

Based on the concept of the rule-based system (Bengio et al. 2015), we develop our semi-supervised text mining naive collision algorithm. Figure 9.1 gives the graphical representation of the proposed algorithm.

Specifically, the inputs of our proposed algorithm are the training set and test set containing feature vectors of textual risk factor headings. Since the contents of Form 10-K textual disclosures tend to be more boilerplate, the same information is presented by similar sentences (Dyer et al. 2017). Thus, we apply the Vector Space Model (VSM) to construct feature vectors by quantifying headings of textual risk disclosures, which can make headings with the same information have higher similarities (Turney and Pantel 2010). In addition, by analyzing the textual risk disclosures, we found that the meaning of a heading is expressed mainly by using nouns, so we select all nouns of a heading to form the feature vector. To implement our algorithm, feature vectors of textual risk factor headings need to be divided into a

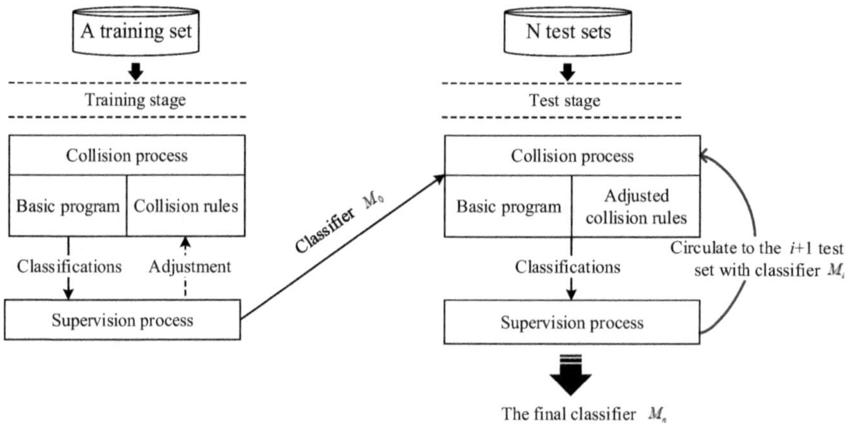

Fig. 9.1 Graphical model of the proposed semi-supervised text mining naive collision algorithm

training set and several test sets. Here, we assume that the feature vectors are divided into one training set and n test sets to illustrate our algorithm.

Having determined the training set and the test sets, we input them into the semi-supervised text mining naive collision algorithm. It is clear from Fig. 9.1 that the algorithm is divided into two stages, and every stage is composed of two processes. To be specific, the collision process automatically classifies risk factors. The supervision process adjusts the automatic classifications with appropriate human interpretation. Thus, our proposed semi-supervised algorithm is characterized by combing the automatic collision process and the human supervision process. In the following, we describe the procedure of our proposed algorithm in detail.

First, in the training stage, the training set is inputted into the collision process to obtain the original automatic classifications via the basic program, under which different feature vectors are merged into the same classification when the similarities between them are beyond the set threshold. The specific pseudocode of the basic program is recorded in Table 9.1. Then, the original automatic classifications are analyzed in the following supervision process. The supervision process consists of two subprocesses of supervision. The first subprocess of supervision aims to adjust the collision rules to make the automatic classifications more accurate and convergent through experts' knowledge. Then, we combine the adjusted collision rules with the basic program in the collision process to update the original automatic classifications. We repeat the first subprocess of supervision until the most suitable adjusted rules are developed. Having determined adjusted collision rules, we can get the most accurate and convergent automatic classifications by combing the adjusted collision rules and the basic program in the collision process.

The second subprocess of supervision is to merge the most accurate and convergent automatic classifications with the same topic. The core subprocess of supervision is labeling each classification to reflect its topic based on experts' judgments. Overall,

Table 9.1 The specific pseudo-code of the basic program in the collision process

Algorithm: the basic program of the collision process

procedure TAKE (v_0)	// Initialize the classifier, *M*
$m_{v0} \leftarrow \{v_0\}$	
$M \leftarrow \{m_{v0}\}$	
end procedure	
procedure COLLISION (V, M, ε)	// Input the data set V, the initialized classifier M and
for $v_i \in V$ **do**	Set the threshold value ε to collide vectors
for m $\in M$ **do**	into the classifier
for $v_j \in m$ **do**	
$similarity_{i,j} \leftarrow \cos(v_i, v_j)$	// Compute the similarity of the vector and vectors
end for	In the classifier
end for	
$similarity_{i,p} \leftarrow \max(similarity_{i,j})$	// Select the maximal similarity
if $similarity_{i,p} \geq \varepsilon$ **then**	// Cluster according to the comparison of the
$m_{v_p} \leftarrow m_{v_p} \cup \{v_i\}$	maximal similarity and the threshold value
Else	
$m_{v_i} \leftarrow \{v_i\}$	
end if	
end for	
end procedure	

in the training stage, we obtain an original classifier, M_0, which contains the risk factor classification results and their corresponding feature vectors.

The original classifier M_0 obtained from the training stage is used as the input of the test stage. Feature vectors contained in the first test set are classified based on the classifier M_0 through the collision process by combing the basic program and adjusted collision rules. Then, the derived classifications are merged to arrive at the appropriate classifier M_1 through the supervision process. The enhanced classifier M_1 is further applied to the next test set to classify feature vectors. In this way, the classifier M_i is used as the input of the $i + 1$ test set to obtain the classifier M_{i+1}. We repeat the collision and supervision processes n times in the test stage until all feature vectors contained in n text sets are gradually classified. Finally, we obtain the most enhanced classifier M_n, which provides the final risk factor classification results and the corresponding feature vector of each risk factor type.

In summary, training and test sets containing feature vectors obtained by VSM are used as inputs of our proposed semi-supervised text mining naive collision algorithm.

Following a training stage and a text stage, we eventually obtain the output of risk factor classifications. The characteristic of semi-supervision is reflected by an automated collision process and a human supervision process in each stage. Compared with the unsupervised Sent-LDA text analysis approach, it provides higher accuracy classifiers in identifying risk factors.

9.3 Data Description

Although the SEC required all filers to discuss company risk factors as one additional section (section 1A) in their Form 10-K in 2005 (SEC 2005), the effectiveness of risk factor disclosures had been widely questioned. In 2010, the SEC issued comment letters asking filers to only include specific risk factors related to the individual company to enhance effectiveness (Zhu et al. 2016; SEC 2010). Thus, to obtain the more effective textual risk disclosures, we extract the summary headings of textual risk disclosures reported in section 1A of Form 10-K filings for 2010–2016 to prepare our data set.

To analyze bank risk factors, the U.S. commercial banks are selected based on an industrial classification (SIC) code list on the SEC's website (https://www.sec.gov/info/edgar/siccodes.htm). The SICs related to commercial banks are 6021, 6022, and 6029, whose corresponding industry names are national commercial banks, state commercial banks, and commercial banks, NEC, respectively. Then, we can collect the names of U.S. commercial banks and their corresponding Form 10-K filings from the EDGAR databases on the SEC's website.

To extract headings from textual risk disclosures and structure them as Excel files, we set rules based on the characteristics that textual risk factor headings are usually marked in italics, boldface, or underlines in the Form 10-K. To verify the accuracy of data extraction based on our rules, we manually checked the structural headings. The results show that only sixty-four out of three thousand headings are misextracted, and the accuracy rate is as high as 97.87%, indicating the robustness of our extraction rules. Eventually, we obtained 59,418 headings of textual risk disclosures from 2189 Form 10-K filings of U.S. commercial banks (27.14 headings per Form 10-K on average) over the period of 2010–2016.

9.4 Empirical Results

In this section, the proposed semi-supervised text mining naive collision algorithm is implemented to discover bank risk factors by analyzing the textual risk disclosures reported in section 1A of U.S. commercial banks' SEC Form 10-K filings. We design the adjusted collision rules according to the characteristics of U.S. commercial banks' textual risk disclosures and verify the convergence of the algorithm. Then, we discuss the classification results of bank risk factors and draw conclusions. Finally, we compare our proposed semi-supervised text mining naive collision algorithm with the competing unsupervised and supervised methods in terms of predictive power and clustering quality.

9.4.1 Algorithm Implementation

Having collected textual risk factor headings, first, we use VSM to quantify them to form feature vectors. To identify annual bank risk factors, the feature vectors are divided into seven sets by year, which allows us to analyze the annual change of bank risk factors over the sample period. Specifically, feature vectors of 2010 are put into a training set. Six test sets contain feature vectors from 2011 to 2016.

After putting the training set into the training stage of our proposed algorithm, we set a higher threshold value of similarity $\varepsilon = 0.71$ under the basic program. The reasons why we set such a threshold are as follows. First, from the perspective of theory, a threshold of similarity larger than $\sqrt{2}/2$ can avoid the problem of over-clustering. For example, α, β, γ are feature vectors made up of nouns. w_1, w_2 are nouns that make up these three feature vectors. The numbers of w_1, w_2 represent the times of w_1, w_2 appearing in a heading of textual risk disclosures, respectively. When $\alpha = (w_1 = 1, w_2 = 1)$, $\beta = (w_1 = 1, w_2 = 0)$, and $\gamma = (w_1 = 0, w_2 = 1)$, we can calculate that $similariy(\alpha, \beta) = \sqrt{2}/2$, $similariy(\alpha, \gamma) = \sqrt{2}/2$, and $similariy(\beta, \gamma) = 0$ based on the formula of similarity calculation. If the similarity threshold is smaller than $\sqrt{2}/2$, α and β will be classified together, and α and γ will be classified together. Hence, α, β, γ will be classified into one classification, which leads to the problem that β and γ with the similarity of 0 are also classified into one classification. Thus, the threshold of similarity should be larger than $\sqrt{2}/2$ to avoid the problem of over-clustering ($\sqrt{2}/2 \approx 0.707$).

Second, from an empirical point of view, many pre-experiments were done to determine the similarity threshold. As we all know, the higher the threshold of similarity, the higher the accuracy of automatic classification results. However, more human efforts are needed to merge a large number of automatic classification results (Turney and Pantel 2010). Thus, the setting of the similarity threshold should balance the classification accuracy and human efforts. The results of pre-experiments show that by taking 0.71 as the threshold of similarity, we can obtain a relatively higher accuracy of automatic classification results. Besides, the workload of the manual merging of classification results is also acceptable. Thus, the similarity threshold set by us is 0.71, which is larger than $\sqrt{2}/2$ ($\sqrt{2}/2 \approx 0.707$) and achieves a better balance between classification accuracy and human work.

Setting the similarity threshold as 0.71, we obtain an original automatic classifier with 2491 bank risk factor classifications through the collision process. Then, in the first sub-supervision process, we check over feature vectors of each classification contained in the automatic classifier to set adjusted collision rules and thus improve the accuracy of classifications. By repeating the above collision process and the first sub-supervision process three times, we found four problems, for each of which we

Table 9.2 The adjusted collision rules for improving the accuracy of classifications

Adjusted collision rules
a. Write down the names of companies, some special symbols, and common words into the stop-words file. Common words are words appearing in many headings while are not representative of bank risk factors, such as *risk*, *company,* and so on
b. Transfer some informative verbs and adjectives (e.g. compete and technological) into corresponding nouns
c. To generate more convergent classification results, if the dimension of the feature vector v_i is equal to two or three, reduce the dimension of v_j to the same dimension of v_i when computing $similarity_{i,j}$, where j varies from 1 to $i - 1$
d. Due to the setting of rule c, the higher dimension v_i is capable of capturing lower dimension vectors. The misextracted text almost has a higher dimension, which leads to less accuracy of classification results. To address this problem, if the dimension of v_i is greater than or equal to nine, regard it as dirty data and drop it

set a corresponding collision rule to address. The four adjusted collision rules are summarized in Table 9.2.

First, we found that some common words contained in the feature vectors, such as *risk* and *company*, are not representative of bank risks. Additionally, the company names and some special symbols are also not representatives of bank risks. Thus, we record these words, including company names, special symbols, and common words in a stop-words file and extract them from the feature vectors. The corresponding rule is the rule "a" recorded in Table 9.2, whose aim is to extract words that are not representative of bank risks from the inputted feature vectors.

Second, for the automatic classifications that are not satisfactory, we check the original summary headings of textual risk disclosures. We find that verbs and adjectives in these headings are informative for explaining the meanings of headings. However, as discussed in Sect. 9.2, we select all nouns of a heading to form the feature vector as the meaning of a heading is expressed mainly using nouns by analyzing the headings of textual risk disclosures. Thus, the informative verbs and adjectives are extracted from feature vectors, which lead to worse automatic classifications. To overcome this problem, we transfer informative verbs and adjectives (e.g., compete and technological) into corresponding nouns (rule "b"). By doing this, we can obtain better automatic classification results.

Third, by analyzing the feature vectors, we find that the number of nouns contained in each feature vector ranges from two to eight in most cases. For the feature vectors containing two or three nouns, it is almost impossible to merge them with other feature vectors that contain more than 3 nouns according to the basic program. This may lead to a worse convergence of automatic classification results with too many classifications. Thus, to generate more convergent classification results, we set rule "c", according to which if the dimension of the feature vector v_i is equal to two or three, reduce the dimension of v_j to the same dimension of v_i when computing $similarity_{i,j}$, where j varies from 1 to $i - 1$.

Last, we set rule "d" to address the problem of the ability of misextracted text to capture headings of textual risk disclosures being too strong. In Sect. 9.3, we show that there are minimal proportions of text extracted from the Form 10-K filings that are misextracted, which is quite common when extracting specific text from a file. For example, Bao and Datta (2014) had some misextracted content. By analyzing the misextracted text, we find that the misextracted text is usually a paragraph or a long sentence, so the corresponding feature vector has many nouns. Due to the setting of rule "c", the higher dimension of the feature vector v_j can capture feature vectors with lower dimensions. Thus, many headings of textual risk disclosures may be clustered together with the misextracted text. Hence, to guarantee the accuracy of the classification, if the dimension of v_j is greater than or equal to nine, v_j is probably formed by the misextracted text, so we regard it as dirty data and drop it.

Based on the basic program and the adjusted collision rules in the collision process, we obtain the original classifier M_0 with 1796 bank risk factor classifications and their corresponding feature vectors. Then, in the second sub-supervision process, we manually label classification results using the human experts' domain knowledge. There are some automatic labeling methods (Mei et al. 2007), but these are not suitable when such labeling requires domain knowledge. Actually, in most topic model research, it is customary to manually label topics, ensuring high labeling quality (Chang et al. 2009). Huang and Li (2011) design a manual labeling procedure that makes use of human experts' domain knowledge. Bao and Datta (2014) adopt the manual labeling procedure designed by Huang and Li (2011) to label risk factors. Thus, here we also use the manual labeling procedure designed by Huang and Li (2011) to give automatic classification results meaningful names.

Specifically, four persons on our research team undertook the work of labeling these 1796 automatic classifications. Their research field is bank risk management; therefore, they have domain knowledge for giving meaningful label names to categorize the corresponding classification results. Besides, they have been trained on a number of real labeled examples of each risk factor before labeling. Each person labels 898 out of 1796 classifications and each classification is labeled by two persons. Thus, based on the four-person domain knowledge of bank risk, we manually label the 1796 classifications with meaningful names. Furthermore, during the process of manual labeling, we found that the classification results are representative of risk factors and are thus easier to label. For each half of the 1796 classifications, the two persons reach a relatively high degree of agreement on labels, which suggests that the classification results obtained through the collision process of our proposed semi-supervised text mining naive collision algorithm are satisfactory. Finally, after merging classifications with the same label, the 1796 original automatic classifications are merged into 20 types of bank risk factors with summary labels.

In the test stage, holding the similarity threshold value constant, we use the original classifier M_0 obtained from the preceding training stage to classify feature vectors contained in the following year 2011 test set. Through the collision process and supervision process, we obtain an enhanced classifier M_1. We then apply M_1 to the 2012 test set to obtain a more enhanced classifier M_2. We repeat the above steps until the last 2016 test set is classified to generate the most enhanced final

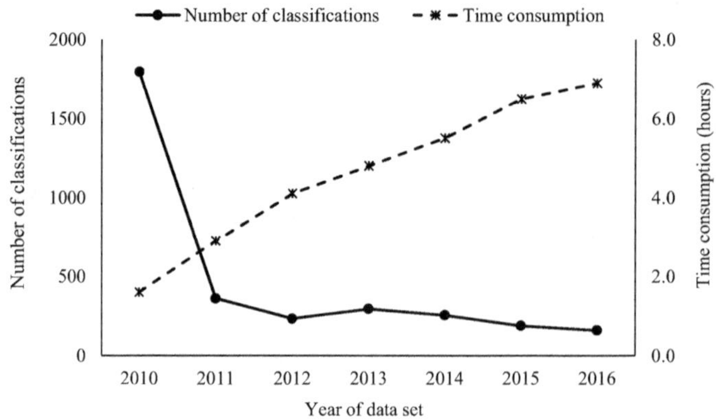

Fig. 9.2 Evolution of classification results and time consumption in the collision process

classifier M_6, which contains the final classification result of 21 bank risk factors, including the regulation, strategy, management operation, macroeconomic factors, loan loss, asset value fluctuation, capital availability, financial expense, competition, development of borrowers, reputation, product and service, merger and acquisition, accounting standard, financial institutions interaction, political environment, third-party cooperation, liability obligation, disaster, country credit rating and other risk factors..

As the classifier is gradually enhanced with a growing number of feature vectors included in each classification, the classification results of seven data sets (including one training set and six test sets) via the automatic collision process become increasingly convergent with fewer and fewer automatic classifications. At the same time, the automatic collision process needs a longer running time. From Fig. 9.2, it is clear that the number of classifications through the collision process experienced a downward tendency by decreasing from 1796 to 160. In contrast, the time consumption of the automatic classification increased from 1.6 to 6.9 h.

9.4.2 Bank Risk Factors Identification Results

We mentioned in Sect. 9.4.1 that through the classifier, the most enhanced final classifier was generated, and the final classification results of 21 bank risks were obtained.We discover 21 bank risk factors by implementing our proposed semi-supervised text mining naive collision algorithm, including the regulation (Reg), strategy (S), management operation (MO), macroeconomic factors (MF), loan loss (L), asset value fluctuation (AVF), capital availability (CA), financial expense (FE), competition (C), development of borrowers (OID), reputation (Rep), product and service (PS), merger and acquisition (MA), accounting standard (AS), financial institutions interaction (FII), political environment (PE), third-party cooperation (TPC), liability obligation (LO), disaster (D), country credit rating (CCR) and other risk factors (ORF). These 21 bank risk factors are visualized using word clods in Fig. 9.3, where the font size corresponds to the probability of the word occurring in the risk factor type. For each type of bank risk factor, we give a brief description in Table 9.3 along with a specific example.

Fig. 9.3 The word clouds of 21 bank risk factors identified based on our proposed algorithm

Table 9.3 Risk factor categorizations and definitions

	Risk factors	Definitions and examples
1	Accounting standard	The risk factor that related to the requirements of accounting standards and the change of them e.g., "*Changes in accounting standards could materially impact the Corporation's financial statements.*"
2	Merger and acquisition	Risks arise from the process of merger and acquisition e.g., "*We will face risks with respect to any future expansion and acquisitions or mergers*"
3	Asset value fluctuation	Potential losses resulted from asset value fluctuations in market investment activities e.g., "*Declines in asset values may result in impairment charges and may adversely affect the value of the Company s investments, financial performance, and capital*"
4	Capital availability	Risk factor related to the ability to support liquidity through funding or assets liquidity with a fair value e.g., "*Capital resources and liquidity are essential to our businesses and could be negatively impacted by disruptions in our ability to access other sources of funding*"
5	Competition	The fierce competition among commercial banks affects banks' profitability and further affects banks' risk-taking behaviors e.g., "*The corporation operates in a highly competitive industry and market Area*"
6	Disaster	Huge damage made by disasters, such as hurricanes and earthquakes e.g., "*Severe weather, natural disasters or other climate change related matters could significantly impact our business*"
7	Financial expense	Financial expenses influence bank profitability, including tax payment, deposit premium expense to the Federal Deposit Insurance Corporation (FDIC), and so on e.g., "*Increases in FDIC insurance premiums may have a material adverse effect on our results of operations*"
8	Financial institutions interaction	The business interactions between banks and other financial industries made risk contagion possible e.g., "*We may be adversely affected by the soundness of other financial institutions*"

(continued)

Table 9.3 (continued)

	Risk factors	Definitions and examples
9	Liability obligation	Risk factors related to various liability obligations taken by banks, appearing as environment obligation, debt, payment, and so on e.g., "*Our obligation to make interest payments to the trust on the debentures is subordinated to existing liabilities or additional debt we may incur*"
10	Loan loss	Potential credit losses related to the credit qualities and default of borrowers e.g., "*Our consumer loans generally have a higher degree of risk of default than our other loans*"
11	Management operation	This includes key factors that are important for daily robust operation, such as personnel, system, information, internal control, technology, litigation, agency problem, and so on e.g., "*We rely heavily on our management team, and the unexpected loss of key management may adversely affect our operations*"
12	Macroeconomic factor	The changes in macroeconomic conditions, such as interest rate, equity index, foreign exchange have a huge impact on banks' activities and therefore may lead to potential risk losses e.g., "*Our results of operations may be adversely affected by changes in national and/or local economic conditions*"
13	Development of borrowers	The developments of other industries, such as energy, car, agriculture, and real estate have major implications for the banks' deposit, loan and investment business e.g., "*Real estate lending in our core Texas markets involves risks related to a decline in value of commercial and residential real estate*"
14	Political environment	The stability of the political environment is important for a bank's operation. Terrorism and war will lead to huge damage e.g., "*Acts or threats of terrorism and political or military actions by the United States or other governments could adversely affect general economic industry conditions*"
15	Product and service	Popular financial products and services provided by banks are sources of profit e.g., "*New lines of business or new products and services may subject us to additional risks. A failure to successfully manage these risks may have a material adverse effect on our business*"

(continued)

Table 9.3 (continued)

	Risk factors	Definitions and examples
16	Regulation	Regulatory laws, policies, acts affect the risk-taking behavior of banks e.g., "*The company operates in a highly regulated environment and may be adversely impacted by changes in laws and regulations*"
17	Reputation	A good reputation is helpful for profitability while a bad reputation made a bank at high risk e.g., "*An impairment in the carrying value of our goodwill could negatively impact our earnings and capital*"
18	Strategy	Decisions on a bank's future growth, including expansion, market share, internationalization, and so on e.g., "*Our growth strategy includes risks that could have an adverse effect on financial performance*"
19	Third-party cooperation	Risk factors associated with the process of cooperation with peers, upstream enterprises, and downstream enterprises e.g., "*The Corporation is subject to risk from the failure of third party vendors*"
20	Country credit rating	The government credit rating affects financial markets and therefore affects ban risks e.g., "*Recent and/or future U.S. credit downgrades or changes in outlook by major credit rating agencies may have an adverse effect on financial markets, including financial institutions and the financial industry*"
21	Other risk factors	Risk factors cannot be categorized into any of these 20 risk factor types are labeled with "other risk factors" e.g., "*Risk factors that may affect future results*"

Several previous studies also analyzed risk factor disclosures to classify risk factors reported in Form 10-K. They use Form 10-K filings from all industries, thereby the identified risk factors are general and exposed by various industries. Unlike previous studies, we collect risk factor disclosures only from commercial banks' Form 10-K filings to discover bank risk factors. Thus, here we make a comparison between our classification results for bank risk factors and general risk factors identified in other studies (summarized in Table 9.4) to find which risk factors are unique to the banking industry.

From Table 9.4, we can see that Huang and Li (2011), Mirakur(2011), Campbell et al (2014), and Bao and Datta (2014) have found various general risk factors and labeled them based on their judgments. Risk factors with the same content may be labeled with different names in different studies. Thus, we compare the bank risk factors discovered by us with general risk factors identified by other researchers

Table 9.4 The summary of risk factor classifications in four previous studies

Risk factor	Studies			
	Huang and Li (2011)	Mirakur (2011)	Campbell et al (2014)	Bao and Datta (2014)
1	Financial conditin risks	Accounting	Financial	Human resources
2	Restructure risks	Acquisitions	Other idiosyncratic	Intellectual property licensing
3	Funding risks	Calamities	Legal and regulatory	Product defects lawsuits
4	Merger and Acquisition risk	Capital expenditures	Other Systematic	Regulation changes
5	Regulation risks	Capital structure	Tax	Catastrophes input prices
6	Catastrophes risks	Cash		Volatile stock price
7	Shareholder's interest risks	Competition		Shareholder's interest
8	Macroeconomic risks	Contracts		Macroeconomic cyclical industry
9	International risks	Credit risk		Cost risks
10	Intellectual property risks	Customer concentration		Rely on large customers
11	Potential defects in products	Distribution		Competition
12	Potential/ongoing lawsuits	Government		Volatile stock price
13	Infrastructure risks	Industry		Debt risks
14	Disruption of operations	Insurance		Funding
15	Human resource risks	Intellectual property		Financial condition risks
16	Licensing related risks	International		Property
17	Suppliers risks	Inventory		Investment
18	Input prices risks	Investments		Regulation changes
19	Rely on a few large customers	Key personnel		Tax risks
20	Competition risks	Labor		International risks
21	Industry is cyclical	Legal		Credit risks
22	Volatile demand and results	Macro		Volatile demand product introduction

(continued)

Table 9.4 (continued)

Risk factor	Studies			
	Huang and Li (2011)	Mirakur (2011)	Campbell et al (2014)	Bao and Datta (2014)
23	Volatile stock price risks	Marketing		Suppliers
24	New product introduction risks	Operations		Accounting risks
25	Downstream risks	Regional		Production introduction
26		Solvency		Downstream
27		Stock price		Infrastructure
28		Suppliers		Credit risks
29		Takeovers		Acquisition restructure
30				Infrastructure operation disruption

in terms of content. We find that 17 of 21 identified bank risk factors are general and can affect the risks of various industries. Only 4 risk factors are unique to the banking industry. In other words, the four risk factors identified by us, including financial institutions interaction, loan loss, development of borrowers, and country credit rating, are only experienced by commercial banks.

9.4.3 Bank Risk Factors Analysis

Having discovered 21 bank risk factors, we further analyze which risk factors have stronger effects on bank risks. In estimating bank risks, selecting several important risk factors instead of putting all risk factors into models will make bank risk measurement more efficient.

9.4.3.1 Top Important Bank Risk Factors

The more frequent the disclosure of a risk factor, the more textual risk factor headings are classified into the risk factor, and the more attention paid by commercial banks to the risk factor, which further indicates that the risk factor is more important for commercial banks. Thus, the importance ratio, which is used to measure the importance of each risk factor, is calculated based on the principle that the annual disclosure frequency of each risk factor can reflect the importance of the risk factor. The annual disclosure frequency of a risk factor is the number of headings classified into the risk factor of each sample year. Thus, we calculate the importance ratio by dividing the number of textual risk factor headings classified into a risk factor by the total number of textual risk factor headings. Formally,

$$\text{Important ratio}_{i,t} = \frac{F_{i,t}}{TF_t} = \frac{F_{i,t}}{\sum\limits_{i=1}^{21} F_{i,t}} \tag{9.1}$$

where Important ratio$_{i,t}$ and $F_{i,t}$ stand for the importance ratio and the disclosure frequency of bank risk factor i at year t, respectively. TF_t represents the total disclosure frequency of all 21 bank risk factors at year t, which is the sum of annual disclosure frequencies of 21 bank risk factors at year t.

In addition to the annual importance ratio, we can also calculate the total importance ratio of a risk factor during a period, representing the importance of the risk factor during the sample period. The total importance ratio is written as

$$\text{Total important ratio}_i = \frac{\sum_{t=1}^{T} F_{i,t}}{\sum_{t=1}^{T} TF} = \frac{\sum_{t=1}^{T} F_{i,t}}{\sum_{t=1}^{T} \sum_{i=1}^{21} F_{i,t}} \tag{9.2}$$

where Total important ratio$_i$ stands for the total importance ratio of the risk factor i during the sample T period.

Overall, the importance ratio is used to measure the importance of each risk factor. The larger value of the importance ratio corresponds to the higher level of importance of the risk factor and vice versa. The calculated total importance ratios of these 21 bank risk factors over the sample period of 2010–2016 are summarized in Table 9.5.

We sort the 21 identified bank risk factors from high to low values of importance ratios in Table 9.5. To select the top important bank risk factors, we take the cumulative importance ratio of 80% as the boundary. It is clear that the cumulative importance ratio of the top 8 of the 21 bank risk factors, including regulation (16.64%), strategy (14.89%), management operation (13.02%), macroeconomic factor (10.91%), loan loss (9.38%), asset value fluctuation (7.22%), capital availability (6.01%) and financial expense (3.21%), account for over 80% (81.28%), which indicates that these eight bank risk factors are essential for bank risk measurement. The cumulative importance of the remaining 13 bank risk factors only accounts for less than 20%. In other words, although the number of important risk factors accounts for only 38.10% of all risk factors, their cumulative importance is up to 81.28%. Using these top 8 important risk factors in bank risk modeling can explain the majority of risk P&L, which makes bank risk measurement more efficient. Of course, researchers can set the boundary of the cumulative importance ratio based on their researches to select important risk factors for measuring bank risks.

Besides, from the top 8 important risk factors, we can see that the top 3 important risk factors, i.e. regulation, strategy, and management operation whose cumulative importance (44.55%) approaching 50% are risk factors affecting non-financial risks. The risk factors affecting financial risks (i.e. credit and market risks), including the

Table 9.5 Importance ratios of 21 bank risk factors based on all samples from 2010 to 2016

Risk factor	Importance ratio (%)	Cumulative importance ratio (%)
Regulation	16.64	16.64
Strategy	14.89	31.53
Management operation	13.02	44.55
Macroeconomic factor	10.91	55.46
Loan loss	9.38	64.84
Asset value fluctuation	7.22	72.06
Capital availability	6.01	78.07
Financial expense	3.21	81.29
Competition	2.73	84.02
Development of borrowers	2.71	86.72
Reputation	2.32	89.04
Product and service	1.93	90.97
Merger and acquisition	1.58	92.54
Accounting standard	1.40	93.95
Other risk factors	1.39	95.33
Financial institutions interaction	1.38	96.71
Political environment	1.23	97.94
Third-party cooperation	1.06	99.00
Liability obligation	0.48	99.48
Disaster	0.36	99.84
Country credit rating	0.16	100.00

macroeconomic factor, loan loss, asset value fluctuation and capital availability, are less important. However, researchers and regulators take credit and market risks as two main risk types faced by banks (BCBS 2006; Li et al. 2015). Therefore, by analyzing the textual risk disclosures, we found that bankers are aware of the importance of non-financial risks. In the future, more attention should be paid to the management of non-financial risks.

Furthermore, we want to study whether the important risk factors for each year are the same. The annual importance ratio can reflect the annual importance of a risk factor. Thus, according to (Eq. 9.1), we calculate the annual importance ratios of 21 bank risk factors from 2010 to 2016. We still take the cumulative importance ratio of 80% as the boundary to select the top important bank risk factors. The annual top 8 important risk factors from 2010 to 2016 are summarized in Table 9.6 for comparison.

From Table 9.6, it is clear that regulation, management operation, macroeconomic factors, loan loss, asset value fluctuation, and capital availability were always important risk factors for commercial banks from 2010 to 2016. It is noteworthy that the importance of strategy increased from out of the top 8 to rank 6, rank 3, and rank 1, which indicates that strategy is increasingly important for commercial banks' risk profiles from 2010 to 2016. In contrast, the development of borrowers became less important and fell out of the top 8 except in 2010. Additionally, from 2010 to 2014, the risk factor type of financial expense ranked approximately 8. However, it was

Table 9.6 Annual rankings of top 8 important bank risk factors from 2010 to 2016

Ranking?	2010	2011	2012	2013	2014	2015	2016
Regulation	1 (15.03%)	1 (18.44%)	1 (16.54%)	2 (16.44%)	2 (17.85%)	2 (16.05%)	2 (16.12%)
Management operation	2 (13.88%)	2 (13.76%)	2 (13.31%)	3 (13.03%)	3 (12.55%)	3 (12.47%)	3 (11.91%)
Macroeconomic factor	3 (13.67%)	3 (11.82%)	4 (11.19%)	4 (10.31%)	4 (9.46%)	4 (9.71%)	4 (9.71%)
Loan loss	4 (11.70%)	4 (10.36%)	5 (9.23%)	5 (9.45%)	5 (8.72%)	5 (7.94%)	5 (7.80%)
Asset value fluctuation	5 (10.97%)	5 (8.47%)	6 (6.40%)	6 (6.19%)	6 (5.89%)	6 (5.93%)	6 (5.94%)
Capital availability	6 (7.79%)	7 (6.34%)	7 (6.02%)	7 (5.73%)	7 (5.75%)	7 (5.14%)	7 (5.00%)
Financial expense	7 (4.17%)	8 (3.83%)	8 (3.75%)	8 (2.93%)	8 (2.64%)	–	–
Development of borrowers	8 (4.02%)	–	–	–	–	–	–
Strategy	–	6 (7.22%)	3 (12.45%)	1 (17.43%)	1 (19.90%)	1 (23.37%)	1 (23.50%)
Reputation	–	–	–	–	–	8 (2.59%)	8 (2.59%)

Note The value in the parenthesis represents the important ratio of the corresponding bank risk factor type. The value above the parenthesis represents the ranking of the important ratio of the corresponding bank risk factor type each year. Other than 2014, in which the top 8 important risk factors account for 78.89%, the top 8 important risk factors in other years account for more than 80%

replaced by reputation in 2015 and 2016, which suggests that the importance of reputation gradually outperformed that of the financial expense.

9.4.3.2 The Change in the Importance of Bank Risk Factors

By analyzing whether the important risk factors of each year are the same, we find that the annual importance of risk factors changed over the sample period. Thus, we further analyze the change of the importance of each risk factor type over the period 2010–2016. As discussed in Sect. 9.4.3.1, the number of risk factor headings classified into a risk factor can reflect the attention paid by commercial banks to the risk factor. Thus, we analyze the change in the importance of each risk factor by comparing the number of risk factor headings classified into a risk factor from 2010 to 2016.

Based on the number of risk factor headings classified into a risk factor summarized in Table 9.7, we can find whether some risk factors appeared or disappeared in later years. The number of headings equals 0 indicates that the corresponding risk factor cannot be identified in this year. We can see from Table 9.7 that the number of headings classified into the risk factor "Country credit rating" is zero in 2010, which indicates that "Country credit rating" appeared in 2011. Additionally, except that the number of headings classified into the risk factor "Country credit rating" is zero in 2010, no other number is zero, which indicates that no other risk factors appeared or disappeared in later years.

Furthermore, we also analyze the trends of the importance of bank risk factors over the sample period according to the number of headings classified into a risk factor summarized in Table 9.7. Specifically, these 21 bank risk factors can be divided into three categories of more important, stable, and less important. The specific change trends of the importance of bank risk factors in these three categories are visually shown in Figs. 9.4, 9.5, and 9.6, respectively.

From Fig. 9.4, it is easy to see that the importance of strategy, political environment, and reputation experienced an increasing tendency. As recorded in Table 9.7, the number of risk factor headings classified into strategy, political environment, and reputation increased from 287, 38, and 116 to 1904, 114, and 210, respectively. The sharp increase in strategy made it become the top 2 important bank risk factor types for commercial banks. Thus, besides top important risk factors, researchers and bankers should pay more attention to these three risk factors that have become increasingly important.

Bank risk factors whose significances were relatively stable from 2010 to 2016 are illustrated in Fig. 9.5. We can see that the changes in the importance of the eight risk factors, including product and service, other risk factors, disaster, regulation, third-party cooperation, country credit rating, merger and acquisition, and accounting standard were slight, without an obvious upward or downward trend. Therefore, the significance levels of these eight risk factors were relatively stable throughout 2010–2016 for commercial banks.

Table 9.7 Number of risk factor headings classified into each of the 21 risk factors from 2010 to 2016

Category	Risk factor	2010	2011	2012	2013	2014	2015	2016
		Number of headings						
More important	Strategy	287	661	986	1363	1616	1877	1904
	Political environment	38	98	145	127	101	93	114
	Reputation	116	203	219	188	211	208	210
Stable	Product and service	135	229	165	150	142	143	160
	Accounting standard	130	103	98	122	116	122	128
	Merger and acquisition	144	126	128	125	121	133	143
	Country credit rating	0	15	15	19	17	12	16
	Third party cooperation	113	94	75	82	82	73	102
	Regulation	1387	1688	1310	1285	1450	1289	1306
	Disaster	43	33	25	33	24	25	26
	Other risk factors	117	114	147	121	110	109	91
Less important	Asset value fluctuation	1013	775	507	484	478	476	481
	Capital availability	719	580	477	448	467	413	405
	Loan loss	1080	948	731	739	708	638	632
	Management operation	1281	1260	1054	1019	1019	1002	965
	Macroeconomic factor	1262	1082	886	806	768	780	787
	Competition	343	286	231	187	201	178	167
	Financial expense	385	351	297	229	214	195	205
	Financial institutions interaction	174	178	169	80	76	60	70
	Liability obligation	93	33	26	28	31	38	29
	Development of borrowers	371	297	229	183	169	169	162

As shown in Fig. 9.6, ten bank risk factors experienced a decreasing tendency in importance from 2010 to 2016. Specifically, from Table 9.7, the number of risk factor headings classified into asset value fluctuation, capital availability, loan loss, management operation, macroeconomic factor, competition financial expense, financial institutions interaction, liability obligation and development of borrowers declined from 1013, 719, 1080, 1281, 1262, 343, 385, 174, 93 and 371 to 481, 405, 632, 965, 787, 167, 205, 70, 29 and 162, respectively. These downward trends represent that the attention paid by U.S. commercial banks to these ten risk factors gradually decreased,

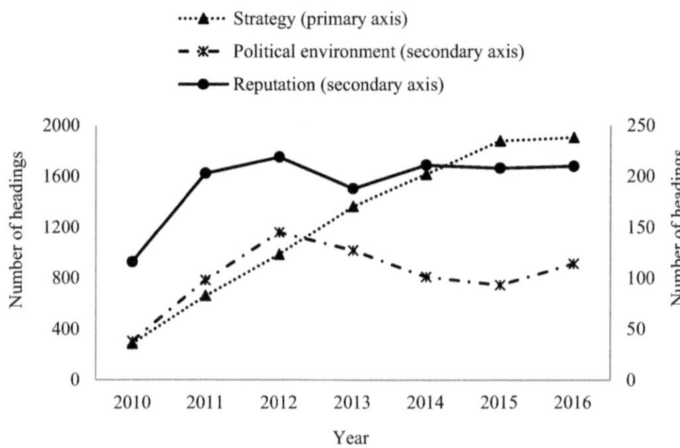

Fig. 9.4 Three risk factors that became more important from 2010 to 2016

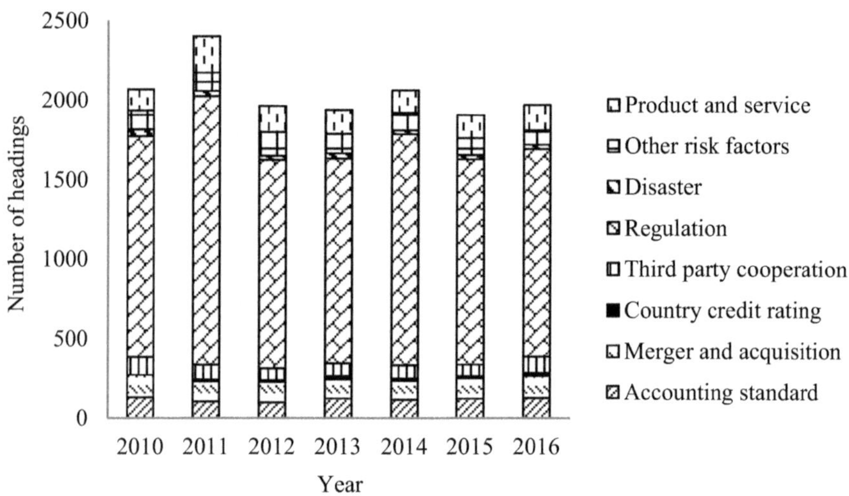

Fig. 9.5 Eight risk factors whose importance keep stable from 2010 to 2016

which further indicates that these ten bank risk factors became less important from 2010 to 2016.

Overall, the importance of each bank risk factor is changing over time. Strategy, political environment, and reputation are the three bank risk factors becoming increasingly important. Basel suggests that banks' business activities inevitably produce various types of risk, including credit, market, operational, liquidity, legal, strategic, country, and reputational risk (BCBS 2006). However, the key issues of bank risk management are the measurements of credit, market, and operational risks (Li et al. 2015), which means the measurements for increasingly important strategic,

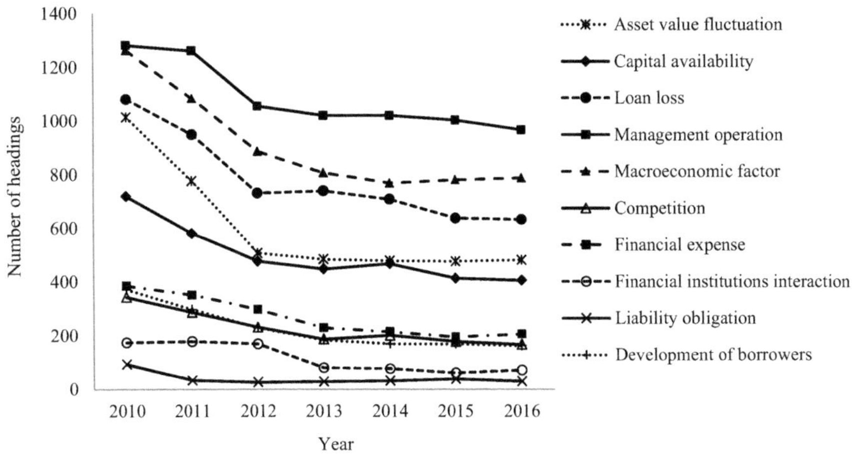

Fig. 9.6 Ten risk factors that experienced a declining trend in importance from 2010 to 2016

political and reputational risks have not been paid enough attention. Thus, in the future, researchers need to put more emphasis on managing strategic, political, and reputational risks.

9.4.4 Results Comparison

To show that our proposed semi-supervised text mining naive collision algorithm can result in more accurate classifications of bank risk factors, we made an empirical comparison between our proposed algorithm and the unsupervised Sent-LDA text mining approach in terms of clustering quality and predictive power (Bao and Datta 2014). The predictive power and clustering quality are evaluated using two commonly used measures of precision and recall (Powers 2011).

Since random sampling is most appropriate for obtaining a representative sample (Grimmer and Stewart 2013), we randomly select 3000 of a total of 59,418 textual risk factor headings for labeling, which is in accordance with Bao and Datta (2014) who made an empirical comparison between Sent-LDA and LDA. As discussed in Sect. 9.4.1, four persons on our research team manually label these 3000 headings with meaningful names. Specifically, each person labels 1500 out of 3000 headings and each heading is labeled by two people. To ensure consistency, we only retain the headings whose labels are agreed upon by both annotators. This leads to a set of 2653 examples. Then, the headings with the same label will be merged as one classification. Thus, by manually labeling the headings of textual risk disclosures, we obtain the most accurate classification result of bank risk factors, which is regarded as the benchmark for measuring the classification accuracies of our proposed semi-supervised naive collision and the unsupervised Sent-LDA text mining approaches.

For convenience, the classification results obtained by manually labeling the 2653 headings are named as manual bank risk factors.

Having obtained the manual bank risk factors by determining the manual labels of 2653 sample headings, we then identify the bank risk factors using the proposed semi-supervised naive collision and the unsupervised Sent-LDA text mining approaches, respectively. The classifications obtained via the proposed semi-supervised text mining naive collision algorithm are named semi-supervised naive collision (SSNC) bank risk factors, and the labels given to them are named SSNC labels. The classifications obtained via the Sent-LDA text mining approach are named Sent-LDA bank risk factors, and the labels given to them are named Sent-LDA labels. The SSNC labels and Sent-LDA labels are collectively called the text mining labels for convenience.

By observing the headings classified into one text mining bank risk factor, SSNC or the Sent-LDA bank risk factor, the headings whose manual labels are consistent with the text mining labels indicate that these headings contained in the text mining bank risk factor are correct. Thus, for one SSNC or Sent-LDA bank risk factor, the precision rate is used to reflect the proportion of headings whose manual labels are consistent with the text mining label. Formally,

$$PR_i = \frac{M_i}{N_i} \tag{9.3}$$

where PR_i denotes the precision rate of text mining bank risk factor i. M_i represents the number of headings whose manual labels are consistent with the label of the text mining bank risk factor i in a text mining bank risk factor i. N_i stands for the number of the entire headings classified into the text mining bank risk factor i. The larger PR_i is, the higher the proportion of the correct headings contained in one text mining bank risk factor, and the higher the accuracy of the classification result obtained through the text mining approach will be.

From the perspective of manual labels of headings, headings with the same manual label may be classified into different text mining bank risk factors. Headings are classified into the correct classification when the label of the text mining bank risk factor is consistent with the manual label of headings. Thus, for a manual bank risk factor, the recall rate is used to measure the proportion of headings that are classified into the correct text mining bank risk factor. The recall rate is written as

$$RR_j = \frac{W_j}{K_j} \tag{9.4}$$

where RR_j denotes the recall rate of the manual bank risk factor j. W_j represents the number of headings that are classified into the correct text mining bank risk factor in a manual bank risk factor j. K_j denotes the number of the entire headings contained in the manual bank risk factor j. The larger RR_j is, the higher the proportion of headings that are classified into the correct text mining bank risk factor, and the higher the accuracy of the classification result.

Overall, the two measures of precision and recall can be used to measure the accuracy of the classification results. Through the confusion matrix, we can calculate the precision rate and recall rate for each of the identified 21 bank risk factors according to (Eq. 9.3) and (Eq. 9.4). Then, by averaging the precision rate and recall rate of identified 21 bank risk factors, we obtain the averaged precision rate and averaged recall rate, respectively. The precision rate and recall rate of each bank risk factor are used to reflect the classification accuracy of the corresponding bank risk factor. The averaged precision rate and averaged recall rate reflect the averaged level of classification accuracy using the text mining approach to classify bank risk factors.

Since we want to make an empirical comparison in terms of clustering quality and predictive power, we need to implement two comparison experiments between our proposed semi-supervised text mining approach and the unsupervised Sent-LDA text mining approach. Specifically, the selected 3000 headings are used for comparing the clustering quality. In other words, we want to compare the performance of our proposed semi-supervised text mining approach and the unsupervised Sent-LDA text mining approach in classifying these 2653 headings of textual risk disclosures. The predictive power evaluates how well a model performs when predicting unobserved documents. Thus, by removing the selected 3000 headings from the total 59,418 headings of textual risk disclosures, we obtain the classifiers of the semi-supervised naive collision and the unsupervised Sent-LDA text mining approaches using the remaining 56,418 headings. Then, we apply the SSNC and Sent-LDA classifiers into the selected 3000 headings to compare the predictive power.

The confusion matrixes are summarized in Tables 9.9, 9.10, and 9.11 of Appendix. The semi-supervised naive collision text mining approach produces the same classification results in the two experiments of clustering quality and predictive power; therefore, these two experiments produce the same confusion matrix (Table 9.9). Tables 9.10 and 9.11 record the confusion matrixes of the unsupervised Sent-LDA text-mining approach in the experiments of clustering quality and predictive power, respectively. The calculated precision rate and recall rate are summarized in Table 9.8.

From Table 9.8, both the precision rate and recall rate indicate that our proposed semi-supervised text-mining naive collision approach performs better than the unsupervised Sent-LDA approach in terms of both clustering quality and predictive power. Specifically, when evaluating the clustering quality, the averaged precision rate of our proposed semi-supervised text mining approach is up to 75.41%, which is far greater than that of the unsupervised Sent-LDA text mining approach (19.53%). The averaged recall rate of our proposed semi-supervised text mining approach (76.05%) is also far greater than that of the unsupervised Sent-LDA text mining approach (19.93%). Therefore, the higher averaged precision rate and averaged recall rate show that our proposed semi-supervised text mining approach can obtain more accurate classification results of bank risk factors. In the experiment of evaluating predictive power, compared with the averaged precision rate (26.54%) and averaged recall rate (27.61%) of the unsupervised Sent-LDA text mining approach, the averaged

Table 9.8 The precision and recall rate of the semi-supervised and the unsupervised text mining approaches in terms of clustering quality and predictive power

Bank Risk factors	Semi-supervised		Unsupervised			
	Precision rate	Recall rate	Precision rate		Recall rate	
	Clustering quality/Predictive power (%)	Clustering quality/Predictive power (%)	Clustering quality (%)	Predictive power	Clustering quality	Predictive power (%)
Accounting standard	97.32	93.97	23.42	77.78	22.41	78.45
Merger and acquisition	99.18	90.30	20.87	20.00	17.91	17.16
Asset value fluctuation	89.12	66.84	18.68	38.38	8.67	19.39
Capital availability	91.37	72.57	23.01	21.30	14.86	13.14
Competition	100.00	80.32	33.70	33.95	16.49	29.26
Disaster	100.00	65.08	32.94	60.00	44.44	64.29
Financial expense	75.36	62.28	42.86	48.68	50.30	55.09
Financial institutions interaction	70.79	74.12	22.75	18.29	56.47	55.29
Liability obligation	90.43	82.52	20.51	20.30	23.30	26.21
Loan	63.12	61.38	18.09	16.28	11.72	9.66
Management operation	89.04	69.89	21.78	30.33	11.83	19.89
Macroeconomic factor	83.44	70.79	25.69	19.17	20.79	12.92
Other industry development	74.82	70.27	24.37	25.74	19.59	23.65
Other risk factors	17.02	25.40	7.48	7.29	17.46	11.11

(continued)

Table 9.8 (continued)

Bank Risk factors	Semi-supervised		Unsupervised			
	Precision rate	Recall rate	Precision rate		Recall rate	
	Clustering quality/Predictive power (%)	Clustering quality/Predictive power (%)	Clustering quality (%)	Predictive power	Clustering quality	Predictive power (%)
Political environment	29.49	100.00	3.27	2.25	10.87	4.35
Product and service	48.10	78.35	6.25	12.63	6.19	12.37
Regulation	78.00	64.29	37.96	26.57	22.53	20.88
Reputation	67.88	81.58	12.00	28.28	13.16	24.56
Strategy	45.14	92.86	3.70	9.45	8.57	17.14
Third party cooperation	74.02	96.91	1.82	34.38	2.06	34.02
Country credit rating	100.00	97.30	8.97	6.20	18.92	21.62
Averaged	**75.41**	**76.05**	**19.53**	**26.54**	**19.93%**	**27.16%**

precision rate (75.41%) and averaged recall rate (76.05%) of our proposed semi-supervised text mining approach are far greater. Thus, our proposed semi-supervised text mining approach performs better in terms of predictive power.

Furthermore, the precision rate and recall rate of our proposed semi-supervised text mining approach are the same in two experiments of clustering quality and predictive power, which indicates that our proposed semi-supervised text mining approach can achieve the same highly accurate classification result, whether used for a small sample or a large sample. However, the unsupervised Sent-LDA performs worse with a small sample. The reason is that the classification accuracy of the unsupervised Sent-LDA depends on the size of the sample. The larger the sample size, the more accurate the classifications. Thus, Bao and Datta (2014) obtained a relatively accurate classification result by using the Sent-LDA text mining approach to deal with a large sample with 322,287 headings. However, the sample size is small in our experiments, which may be the reason that the classification result obtained by the Sent-LDA is less precise. In particular, since we obtain the Sent-LDA classifier based on 56,418 headings in the experiment on evaluating predictive power, rather than 3000 headings used in the experiment on clustering quality, the averaged precision rate (26.54%) and averaged recall rate (27.61%) in the experiment on predictive power are larger than the averaged precision rate (19.53%) and the averaged recall rate (19.93%) in the experiment on clustering quality, which indicates that the Sent-LDA performs even worse in the experiment on clustering quality with a smaller sample.

To summarize, the far greater precision rate and recall rate show that our proposed semi-supervised approach performs better in terms of clustering quality and predictive power. Furthermore, the classification accuracy of our proposed semi-supervised text mining approach is not affected by the size of the sample. Even based on a small sample, our proposed semi-supervised text mining approach can produce the same accurate classification result as the large sample.

9.5 Conclusion

The identification of bank risk factors in previous studies mainly depends on the researchers' adjustments or traditional theories, which is infeasible to identify bank risk factors comprehensively and accurately. We found that qualitative textual risk disclosures reported in financial statements contain massive and comprehensive information on bank risks. Thus, the main contribution of this chapter is to comprehensively and accurately discover bank risk factors from textual risk disclosures reported in financial statements of the banking industry.

To comprehensively and accurately discover bank risk factors from textual risk disclosures, we propose a new semi-supervised text mining naive collision algorithm. Compared with the typical unsupervised Sent-LDA text mining approach, we empirically prove that our proposed semi-supervised text mining naive collision algorithm can deliver a higher accuracy classification result of bank risk factors in terms of clustering quality and predictive power. Therefore, we realized the comprehensive and accurate identification of bank risk factors from textual risk disclosures reported in financial statements using our proposed semi-supervised text mining naive collision algorithm.

Specifically, in the experiment, based on 59,418 textual risk factor headings collected from 2189 U.S. commercial banks' Form 10-K filings from 2010 to 2016, we comprehensively identified 21 bank risk factors. We further analyze the importance of each risk factor based on the disclosure frequency. The top 3 important risk factors whose cumulative importance approaching 50% are non-financial risk factors, i.e. regulation, strategy, and management operation. Besides, by analyzing the annual change of the importance of each risk factor, we found that the risk factors of strategy, political environment, and reputation have become increasingly important for commercial banks' risk profiles.

The findings of our empirical study have practical implications for researchers and regulators. First, the 21 bank risk factors identified from textual risk disclosures are far more than those identified in previous studies. Using these 21 bank risk factors for further bank risk measurement can describe the bank risk profiles more comprehensively and accurately. Second, more attention should be paid to non-financial risks faced by banks. Although financial risks, i.e. credit and market risks are traditionally considered as main risk types faced by banks, we found that the top 3 important risk factors identified from textual risk disclosures affect non-financial bank risks. Thus, researchers and regulators should put more emphasis on the management of non-financial risks. Third, as the risk factors of strategic, political, and reputational have become increasingly important, the management of them is of great importance for banks. However, the previous studies mainly focus on the measurement of credit, market, and operational risks. Therefore, there is a need to strengthen the measurement of strategic, political, and reputational risks. To summarize, based on our findings from textual risk disclosures, researchers and regulators can improve bank risk management from the above three aspects.

This chapter is not without limitations. The limitation of identifying bank risk factors from textual risk disclosures is that textual risk disclosures reported in financial statements do not contain all information on corporate risks. In other words, some risk information may not be disclosed by firms in the textual risk disclosures.

Appendix

See Tables 9.9, 9.10 and 9.11.

Table 9.9 The confusion matrix of the semi-supervised text mining naïve collision approach in the two experiments of clustering quality and predictive power

	AS	MA	AVF	CA	C	D	FE	FII	LO	L	MO	MF	OID	ORF	PE	PS	Reg	Rep	S	TPC	CCR
AS	109	0	0	0	0	0	0	0	0	0	0	6	0	0	0	0	0	0	1	0	0
MA	1	121	0	0	0	0	6	0	0	2	3	0	0	0	0	1	0	0	0	0	0
AVF	0	0	131	2	0	0	6	0	0	14	0	1	1	19	2	3	0	13	4	0	0
CA	0	0	2	127	0	0	6	1	2	2	0	0	11	14	0	10	0	0	0	0	0
C	0	0	0	0	151	0	0	0	7	7	0	1	0	4	2	1	7	0	8	0	0
D	0	0	0	0	0	82	0	0	0	1	0	0	7	0	35	1	0	0	0	0	0
FE	0	0	2	0	0	0	104	0	0	0	4	0	0	17	0	37	0	0	0	3	0
FII	0	0	1	0	0	0	0	63	0	3	0	1	1	0	8	0	0	0	8	0	0
LO	0	0	0	0	0	0	0	0	85	0	0	0	0	1	0	0	15	0	2	0	0
L	0	0	1	0	0	0	1	2	0	89	1	2	2	6	0	8	3	0	23	7	0
MO	0	0	0	8	0	0	0	1	0	0	130	1	0	3	1	3	8	25	0	6	0
MF	0	0	0	0	0	0	0	4	0	1	0	126	7	7	22	0	0	0	11	0	0
OID	0	0	0	0	0	0	0	6	0	17	0	7	104	6	0	5	0	0	1	2	0
ORF	1	0	1	0	0	0	6	0	0	0	7	6	4	16	1	13	0	0	7	1	0
PE	0	0	0	0	0	0	0	0	0	0	0	0	0	0	46	0	0	0	0	0	0
PS	0	0	0	0	0	0	1	0	0	0	0	0	1	0	0	76	0	6	0	13	0
Reg	0	1	9	2	0	0	1	4	0	0	0	0	1	1	39	0	117	0	6	1	0
Rep	0	0	0	0	0	0	7	7	0	0	1	0	0	0	0	0	0	93	6	0	0
S	0	0	0	0	0	0	0	0	0	5	0	0	0	0	0	0	0	0	65	0	0
TPC	1	0	0	0	0	0	0	0	0	0	0	0	0	0	0	0	0	0	2	94	0
CCR	0	0	0	0	0	0	0	1	0	0	0	0	0	0	0	0	0	0	0	0	36

Note Acronyms stand for the 21 bank risk factors shown in Sect. 9.4.2

Table 9.10 The confusion matrix of the unsupervised Sent-LDA text mining approach in the experiment of clustering quality

	AS	MA	AVF	CA	C	D	FE	FII	LO	L	MO	MF	OID	ORF	PE	PS	Reg	Rep	S	TPC	CCR
AS	26	0	3	0	0	1	1	0	0	0	0	28	0	0	0	0	0	1	0	36	20
MA	1	24	0	2	0	0	12	1	3	1	5	5	3	0	64	4	0	8	0	1	0
AVF	25	0	17	7	9	0	1	20	0	13	11	7	32	0	1	23	1	9	1	13	6
CA	11	8	0	26	7	7	12	2	0	9	0	19	14	5	8	10	6	0	26	5	0
C	0	18	12	3	31	5	0	40	25	15	6	19	0	3	0	0	0	0	0	10	1
D	3	0	0	2	3	56	0	2	6	0	5	0	1	0	0	0	0	0	48	0	0
FE	0	0	0	0	2	10	84	1	0	17	0	1	4	1	1	7	11	28	0	0	0
FII	3	0	0	2	1	0	0	48	0	1	0	0	4	12	7	0	1	0	0	6	0
LO	11	0	13	0	0	0	0	5	24	0	0	1	1	0	0	10	0	0	10	11	17
L	0	0	0	4	0	0	15	13	2	17	4	2	19	12	0	16	2	6	29	4	0
MO	7	15	18	1	10	25	6	1	1	0	22	0	4	18	13	10	8	10	3	9	5
MF	1	5	0	2	4	0	9	31	8	0	2	37	2	20	16	0	3	28	10	0	0
OID	0	1	7	2	1	0	24	26	11	6	11	0	29	16	0	0	1	0	12	1	0
ORF	1	0	0	5	9	0	2	2	0	2	8	1	0	11	9	0	4	0	6	1	2
PE	0	0	0	0	0	7	11	0	9	0	1	0	0	0	5	0	7	0	0	1	5
PS	0	3	0	0	1	11	8	8	7	0	3	14	1	22	1	6	0	8	2	0	2
Reg	4	1	18	32	1	1	1	1	3	9	5	10	0	14	15	1	41	1	9	6	9
Rep	0	13	2	14	1	9	3	3	0	1	18	0	4	1	5	6	15	15	0	4	0
S	11	14	1	9	1	2	0	0	2	3	0	0	0	1	7	2	6	1	6	0	4
TPC	0	13	0	2	11	36	7	7	7	0	0	0	1	8	1	1	1	0	0	2	0
CCR	7	0	0	0	0	0	0	0	9	0	0	0	0	3	0	0	1	10	0	0	7

Note Acronyms stand for the 21 bank risk factors shown in Sect. 9.4.2

Table 9.11 The confusion matrix of the unsupervised Sent-LDA text mining approach in the experiment of predictive power

	AS	MA	AVF	CA	C	D	FE	FII	LO	L	MO	MF	OID	ORF	PE	PS	Reg	Rep	S	TPC	CCR
AS	91	2	0	0	1	0	0	1	0	0	0	13	0	0	0	0	0	0	0	0	8
MA	1	23	0	2	0	0	49	2	22	0	8	1	8	7	1	1	1	4	0	0	4
AVF	0	0	38	0	14	0	0	0	4	13	45	7	15	16	0	0	0	1	8	12	23
CA	5	0	4	23	15	16	7	1	2	10	1	7	5	1	4	22	0	2	25	2	23
C	0	0	0	0	55	8	6	44	1	0	1	8	0	0	7	17	32	0	0	7	2
D	2	0	5	0	0	81	4	30	0	0	2	0	1	0	0	1	0	0	0	0	0
FE	0	5	1	2	3	4	92	0	0	20	0	9	2	11	0	0	1	0	0	17	0
FII	10	1	5	1	1	0	0	47	0	3	1	2	0	0	0	0	0	8	0	0	6
LO	0	2	0	13	13	0	0	6	27	0	0	11	5	2	2	2	0	10	10	0	0
L	0	14	6	6	21	0	0	2	3	14	0	1	12	8	6	2	20	3	27	0	0
MO	2	0	1	3	10	0	13	10	4	1	37	9	1	23	5	9	8	23	6	10	11
MF	0	0	6	3	0	5	3	69	25	0	7	23	2	0	0	3	22	0	9	0	1
OID	0	0	8	19	1	4	1	14	0	1	0	10	35	0	0	11	5	1	16	0	22
ORF	1	5	0	0	0	0	1	3	13	5	2	4	9	7	1	0	4	3	1	4	0
PE	0	0	0	0	0	9	0	0	8	0	8	0	11	0	2	0	7	0	1	0	0
PS	0	5	14	17	2	0	2	6	8	1	1	3	5	2	13	12	1	0	0	4	1
Reg	1	35	11	10	1	4	9	0	1	12	1	3	6	2	25	4	38	10	2	6	1
Rep	0	10	0	1	23	1	0	11	14	1	1	2	0	10	0	1	0	28	4	1	6
S	0	11	0	6	0	2	0	0	0	5	5	2	8	5	4	1	3	0	12	0	6
TPC	0	2	0	2	2	1	2	1	1	0	2	0	11	2	19	0	1	5	6	33	7
CCR	4	0	0	0	0	0	0	10	0	0	0	5	0	0	0	9	0	1	0	0	8

Note Acronyms stand for the 21 bank risk factors shown in Sect. 9.4.2

References

Aas K, Dimakos XK, Oksendal A (2007) Risk capital aggregation. Risk Manage 9(2):82–107. https://doi.org/10.1057/palgrave.rm.8250027

Alexander C, Peizer J (2003) On the aggregation of market and credit risks. In: ISMA Centre discussion papers in finance No. 2003-13. University of Reading

Alibux A (2007) Cross-border mergers and acquisitions in the European banking sector. Dissertation (Erasmus University of Rotterdam)

Bao Y, Datta A (2014) Simultaneously discovering and quantifying risk types from textual risk disclosures. Manage Sci 60(6):1371–1391. https://doi.org/10.1287/mnsc.2014.1930

Basel Committee on Banking Supervision (2006) International convergence of capital measurement and capital standards: a revised framework. Bank for International Settlements, Basel

Bellini T (2013) Integrated bank risk modeling: a bottom-up statistical framework. Eur J Oper Res 230(2):385–398. https://doi.org/10.1016/j.ejor.2013.04.031

Bengio Y, Goodfellow IJ, Courville A (2015) Deep learning. Nature 521:436–444. https://doi.org/10.1038/nature14539

Berger A, Buch C, Delong G, DeYoung R (2004) Exporting financial institutions management via foreign direct investment mergers and acquisitions. J Int Money Financ 23:333–366. https://doi.org/10.1016/j.jimonfin.2004.01.002

Breuer T, Jandacka M, Rheinberger K, Summer M (2010) Does adding up of economic capital for market and credit risk amount always to conservative risk estimates? J Bank Finance 34(4):703–712. https://doi.org/10.1016/j.jbankfin.2009.03.013

Brewer T, Rivoli P (1990) Politics and perceived country creditworthiness in international banking. J Money, Credit, Bank 22:357–369

Buch C, DeLong G (2004) Cross-border bank mergers: what lures the rare animal? J Bank Finance 28:2077–2102. https://doi.org/10.1016/j.jbankfin.2003.08.002

Campbell JL, Chen HC, Dhaliwal DS, Lu HM, Steele LB (2014) The information content of mandatory risk factor disclosures in corporate filings. Rev Acc Stud 19(1):396–455. https://doi.org/10.1007/s11142-013-9258-3

Chang J, Boyd-Graber JL, Gerrish S, Wang C, Blei DM (2009) Reading tea leaves: how humans interpret topic models. In: Bengio Y, Schuurmans D, Lafferty JD, Williams CKI, Culotta A (eds) Advances in neural information processing systems. Curran Associates, New York, pp 288–296

Chari A, Gupta N (2008) Incumbents and protectionism: the political economy of foreign entry liberalization. J Financ Econ 88:633–656. https://doi.org/10.1016/j.jfineco.2007.07.006

Dyer T, Lang M, Stice-Lawrence L (2017) The evolution of 10-K textual disclosure: evidence from latent dirichlet allocation. J Acc Econ 64(2–3):221–245. https://doi.org/10.1016/j.jacceco.2017.07.002

Embrechts P, Puccetti G, Rüschendorf L (2013) Model uncertainty and VaR aggregation. J Bank Finance 37(8):2750–2764. https://doi.org/10.1016/j.jbankfin.2013.03.014

Gleason K, Mathur I, Wiggins RA (2006) The use of acquisitions and joint ventures by U.S. banks expanding abroad. J Financ Res 29:503–522. https://doi.org/10.1111/j.1475-6803.2006.00191.x

Grimmer J, Stewart BM (2013) Text as data: the promise and pitfalls of automatic content analysis methods for political texts. Polit Anal 21(3):267–297. https://doi.org/10.1093/pan/mps028

Grundke P (2009) Importance sampling for integrated market and credit portfolio models. Eur J Oper Res 194(1):206–226. https://doi.org/10.1016/j.ejor.2007.12.028

Grundke P (2010) Top-down approaches for integrated risk management: how accurate are they? Eur J Oper Res 203(3):662–672. https://doi.org/10.1016/j.ejor.2009.09.015

Hagendorff J, Keasey K (2009) Post-merger strategy and performance: evidence from the US and European banking industries. Acc Finance 49(4):725–751. https://doi.org/10.1111/j.1467-629X.2009.00306.x

Huang KW, Li ZL (2011) A multilabel text classification algorithm for labeling risk factors in SEC form 10-K. ACM Trans Manage Inf Syst (TMIS) 2(3):1–19. https://doi.org/10.1145/2019618.2019624

Jarrow RA, Turnbull SM (2000) The intersection of market and credit risk. J Bank Finance 24(1):271–299. https://doi.org/10.1016/S0378-4266(99)00060-6

Kretzschmar G, McNeil AJ, Kirchner A (2010) Integrated models of capital adequacy—Why banks are undercapitalized. J Bank Finance 34(12):2838–2850. https://doi.org/10.1016/j.jbankfin.2010.02.028

Li J, Guisinger S (1992) The globalization of service multinationals in the "triad" regions: Japan, Western Europe and North America. J Int Bus Stud 23:675–696. https://doi.org/10.1057/palgrave.jibs.8490283

Li J, Zhu X, Lee CF, Wu D, Feng J, Shi Y (2015) On the aggregation of credit, market and operational risks. Rev Quant Financ Acc 44(1):161–189. https://doi.org/10.1007/s11156-013-0426-0

Medova EA, Smith RG (2005) A framework to measure integrated risk. Quant Finance 5(1):105–121. https://doi.org/10.1080/14697680500117583

Mei QZ, Shen XH, Zhai CX (2007) Automatic labeling of multinomial topic models. In: Berkhin P, Caruana R, Wu X (eds) Proceedings of the 13th ACM SIGKDD international conference on Knowledge discovery and data mining. ACM, New York, pp 490–499

Miller GS (2017) Discussion of "the evolution of 10-K textual disclosure: evidence from latent dirichlet allocation." J Account Econ 64(2–3):246–252. https://doi.org/10.1016/j.jacceco.2017.07.004

Mirakur Y (2011) Risk disclosure in SEC corporate filings. In: Working paper. University of Pennsylvania, Philadelphia

Powers DMW (2011) Evaluation: from precision, recall and f-factor to ROC, informedness, markedness & correlation. J Mach Learn Technol 2:2229–3981. https://doi.org/10.9735/2229-3981

Rosenberg JV, Schuermann TA (2006) General approach to integrated risk management with skewed, fat-tailed risks. J Financ Econ 79(3):569–614. https://doi.org/10.1016/j.jfineco.2005.03.001

SEC (2005) Securities and exchange commission final rule, release no. 33-8591(FR-75). Available at: http://www.sec.gov/rules/final/33-8591.pdf

SEC (2010) Annual report pursuant to section 13 or 15(d) of the securities exchange act of 1934, general instructions. Available at: http://www.sec.gov/about/forms/form10-k.pdf

Turney PD, Pantel P (2010) From frequency to meaning: vector space models of semantics. J Artif Intell Res 37:141–188. https://doi.org/10.1613/jair.2934

Zhu X (2007) Semi-supervised learning literature survey. Comput Sci 37(1):63–77. https://doi.org/10.1016/j.patrec.2013.10.008

Zhu X, Yang SY, Moazeni S (2016) Firm risk identification through topic analysis of textual financial disclosures. In: Computational Intelligence. IEEE, pp 1-8. https://doi.org/10.1109/SSCI.2016.7850005

Chapter 10
Bank Risk Aggregation with Forward-Looking Textual Risk Disclosures

10.1 Introduction

The existing bank aggregation approaches based on data from financial statements can be divided into four types. The first financial statement-based bank risk aggregation approach aggregates risks using data from the balance sheet (Kretzschmar et al. 2010; Drehmann et al. 2010; Alessandri and Drehmann 2010). Since profits and losses arising from bank risks within certain periods are recorded in the income statement, researchers have realized that looking for risk proxies from income statements provides a reasonable basis for decomposing profits and losses into different risk sources. Thus, the second financial statement-based bank aggregation approach aggregates bank risks based on first-hand risk profit and loss (P&L) data by mapping income statements' P&L items into different risk types (Kuritzkes and Schuermann 2007).

The former two approaches are based on risk data collected from one piece of financial statements, either income statements or balance sheets. The third financial statement-based bank aggregation approach aggregates individual risks using data drawn from both balance sheets and income statements (Rosenberg and Schuermann 2006). However, the usual practice in bank risk aggregation ignores the off-balance-sheet (OBS) items. Rosenberg and Schuermann (2006) have realized that OBS items can be influential, and the aggregation results may be somewhat arbitrary without considering OBS items. Thus, the last type of financial statement-based approach incorporates OBS items into bank risk aggregation (Li et al. 2018).

To summarize, financial statement-based bank risk aggregation approaches proposed in previous studies have made full use of numerical data recorded in financial statements, including on-balance and off-balance sheets, and income statements. However, the numerical data recorded in financial statements primarily summarize a bank's historical operating performance, which has the main drawback of hysteresis (Beneish et al. 2015). Backward-looking numerical financial statement data used for bank risk aggregation can only describe historical bank risk profiles and do not

J. Li et al., *Financial Statements-Based Bank Risk Aggregation*, Innovation in Risk Analysis, https://doi.org/10.1007/978-981-19-0408-0_10

reflect future conditions of bank risk. Changes in macroeconomic or political conditions may make bank risks differ in their historical volatility characteristics (Kupiec and Ramirez 2013; Jiménez et al. 2013). Thus, bank risk aggregation based on historical numerical financial statement data is less timely. The use of historical data while disregarding changes in future market movements in risk aggregation may generate biased aggregation results, which will lead to inadequate capital against potential total losses. So one common weakness of all the financial statements-based bank risk aggregation approaches discussed above is that the future trends of bank risks are ignored based on historical data.

Hence, forward-looking information is needed to be incorporated into bank risk aggregation to overcome the hysteresis of historical numerical financial statement data. Besides backward-looking information, forward-looking information is also disclosed in financial statements (Kılıç and Kuzey, 2018). In particular, the SEC has emphasized that investors had a greater need for forward-looking disclosures than for disclosures about past events, and it has issued guidelines for companies to present any known trends, plans, and uncertainties that are likely to materially affect future operations (Muslu et al. 2014). Recently, in view of the importance of forward-looking information, some studies have begun to analyze the content of forward-looking information disclosed in financial statements (Athanasakou and Hussainey 2014; Abed et al. 2016).

Since 2005, textual risk factors disclosed in Sect. 1A of Securities and Exchange Commission (SEC) Form 10-K have offered forward-looking risk information to discuss "the most significant factors that make the offering speculative or risky" (SEC 2005). Unlike numerical data recorded in financial statements, which primarily summarize banks' historical operating performance, textual risk disclosures include information on risks arising from future market movements. Thus, textual risk disclosures are forward-looking sources of bank risk information that provide a detailed description of bank risk losses that may occur in the future (Zhu et al. 2016).

Recently, textual risk disclosures have become some of the most helpful analysis segments of annual financial statements (Campbell et al. 2014). Some studies have provided insight into analyzing textual risk disclosures (Huang and Li 2011; Dyer et al. 2017). Text mining algorithms are used to mine valuable information from large volumes of unstructured text (Lüdering and Tillmann 2018; Feuerriegel and Gordon 2018). Four main automatic text mining approaches have emerged so far to analyze textual risk disclosures.

Specifically, Campbell et al. (2014) classified risk factors into five types using a predefined dictionary. Huang and Li (2011) proposed a supervised text mining method called the multi-label categorical K-nearest neighbor (ML-CKNN) to categorize 25 risk factors. However, a list of risk factors is needed to be predefined when using the dictionary approach and supervised ML-CKNN approach. In most cases, however, the risk factors might be hard to derive beforehand. Huang and Li (2011) realized that predefining the number of risk factor types based on researchers' subjective judgments will lead to the problem that some important risk factor types may be left out.

Compared with the dictionary and supervised ML-CKNN approaches, the unsupervised Sentence Latent Dirichlet Allocation (Sent-LDA) model proposed by Bao and Datta (2014) can comprehensively discover risk factors from textual risk disclosures rather than predefine a set of categories of risk factors. It has been used by Zhu et al. (2016) and Wei et al. (2019b). As an unsupervised algorithm, however, one weakness of the Sent-LDA is that it makes researchers stay away from data. The absence of appropriate human interpretation and adjustment may lead to a biased classification result (Miller 2017).

The last type of text mining approach is the semi-supervised naive collision algorithm proposed by Wei et al. (2019a), which can comprehensively identify risk factors with a more accurate classification result by using a large amount of unlabeled data, together with labeled data (Zhu, 2017). Wei et al. (2019a) also empirically confirmed that the semi-supervised naive collision algorithm performs better in classification accuracy than the unsupervised Sent-LDA.

Thus, the semi-supervised naive collision algorithm outperforms the dictionary, supervised, and unsupervised text mining approaches. Hence, this chapter adopts the semi-supervised naive collision algorithm to analyze textual risk disclosures reported in Sect. 1A of SEC Form 10-K. This book first incorporates forward-looking textual risk disclosures reported in financial statements into bank risk aggregation methods. Compared to previous studies that aggregate risks based on historical numerical data recorded in financial statements, we address time lags arising from the only use of historical data. Thus, by using forward-looking textual risk disclosures to adjust risk aggregation results based on historical data, we can obtain more reasonable aggregate results to more likely cover future potential total losses. In our experiment, from a dataset covering 153 U.S. commercial banks for the period of 2010–2017, we collect 812 pieces of numerical risk data and 36,178 headings of textual risk disclosures from 1224 Form 10-K filings to aggregate credit, market, and operational risks of the U.S. banking system.

The remainder of this chapter is organized as follows. Section 10.2 presents the procedure used to incorporate forward-looking textual risk disclosures into bank risk aggregation. Section 10.3 describes the data collection. Section 10.4 discusses our main empirical results. Section 10.5 concludes with a summary of our findings.

10.2 Forward-Looking Adjusted Aggregation Approach

The core issue and difficulty of incorporating textual risk disclosures into bank risk aggregation is the quantification of textual risk disclosures. In the following, we discuss how to quantify textual risk disclosures so as to incorporate them into bank risk aggregation.

As is shown in Fig. 10.1, the whole process of incorporating forward-looking textual risk disclosures into bank risk aggregation is divided into four steps. First, bank risk factors are identified using the semi-supervised text mining naive collision algorithm to analyze textual risk disclosures. The identified bank risk factors are

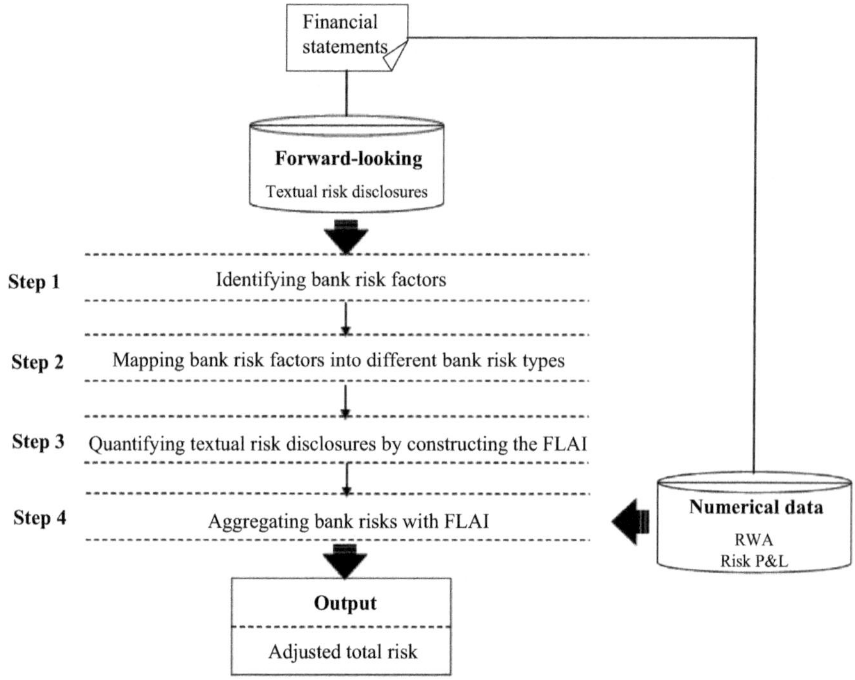

Fig. 10.1 The whole process of incorporating forward-looking textual risk disclosures into bank risk aggregation

then mapped into different types of risks. We then quantify textual risk disclosures by constructing the forward-looking adjustment index (FLAI). Finally, the FLAI is used to enlarge or reduce the total risk values based on historical numerical data, so as to obtain the final forward-looking adjusted total risk. In the following, we describe these four steps in more detail.

10.2.1 Identify Bank Risk Factors Based on Textual Risk Disclosures

Textual risk disclosures reported in Form 10-K consists of a summary heading and detailed explanations, including forward-looking information used to foresee risk factors exposed by banks. Thus, as a first step, we use the semi-supervised naive collision algorithm to identify bank risk factors from textual risk disclosures.

Here, we give an outline of using the semi-supervised text mining naive collision algorithm to identify bank risk factors. The detailed description of the naive collision algorithm is in Sect. 9.2 of Chap. 9. Specifically, inputs are feature vectors, each of which contains all nouns of a summary heading of textual risk disclosures. To identify annual bank risk factors, feature vectors are divided into different sets by year. Then, feature vectors are input into the semi-supervised text mining naive collision algorithm, generating final outputs after experiencing a collision process and a supervision process. Risk factors are automatically classified during the collision process based on the basic program and adjusted collision rules. Once the similarities between different feature vectors exceed the set threshold, they are merged into the same classification. The specific pseudocode and the adjusted collision rules of the basic program, presented in 10.1 and 10.2, are implemented using an open-source software named python (https://www.python.org).

Then, the automatic classification result needs to be adjusted with appropriate human knowledge during the supervision process. Specifically, each automatic classification is labeled to reflect its topic based on expert judgments. Automatic classifications with the same label are merged as a type of risk factor. To summarize, using the semi-supervised text mining naive collision algorithm, we can obtain three outputs: bank risk factor identification results with labels, definitions, and examples; the number of headings classified into each risk factor for each sample year, which is used as the annual disclosure frequency of each risk factor; and the word cloud for each risk factor, in which the font size corresponds to the probability of a word occurring in the risk factor.

10.2.2 Map Bank Risk Factors into Different Types of Risk

As a second step, we map identified bank risk factors into different risk types. In previous studies, through analyses of Form 10-K textual risk disclosures, Huang and Li (2011), Bao and Datta (2014) and Campbell et al. (2014) also identified various risk factors and labeled them based on their judgments. This book further establishes a mapping relationship between identified risk factors and risk types to link the textual risk disclosures into different bank risk types. As individual risk is the basis for further bank risk aggregation, the mapping relationship between risk factors and risk types essentially establishes a link between textual risk disclosures and bank risk aggregation.

Bank risks are affected by certain risk factors (Jarrow and Turnbull 2000; Baselga-Pascual et al. 2015). For example, market risk arises from adverse movements in market factors such as interest rates, exchange rates, and equity prices (Hartmann 2010). Since the changes in risk factors can result in bank losses, some researchers have used risk factors to explain risk loss data (Breuer et al. 2010; Grundke 2010; Rosenberg and Schuermann 2006). Thus, it is feasible and reasonable to establish a mapping relationship between identified bank risk factors and different risk types.

The mapping relationship between bank risk factors and risk types is based on definitions of bank risk factors and bank risks. The definition of each risk factor is obtained from the first stage. Bank risks have been defined by the BCBS and in several previous works (BCBS 2006; Li et al. 2015). Thus, by analyzing definitions of bank risk factors and bank risks, a reasonable relationship between bank risk factors and bank risks can be established. In establishing a mapping relationship between bank risk factors and different risk types, we can determine the annual disclosure frequency and word cloud figure of each type of risk by mapping risk into bank risks factors.

10.2.3 Quantify the Annual Disclosure Frequency of Risk Types by Constructing the FLAI

In the next step, we construct the FLAI based on the annual disclosure frequency of each risk type to quantify textual risk disclosures. Indeed, the more frequently a given type of risk is disclosed, the more attention paid by commercial banks to this risk type, further showing that the risk will become more severe in the forthcoming future. Thus, the FLAI is constructed based on the principle that the annual disclosure frequency of each risk type can reflect foresight on risk severity.

Specifically, for each risk type, we calculate the mean value of annual disclosure frequency for the given time period.

$$M_j = \sum_{t=1}^{T} F_{j,t}/T \tag{10.1}$$

where M_j denotes the mean value of the annual disclosure frequency of risk j $F_{j,t}$ for the sample period T, representing the general condition of foresight on risk j for sample period T.

Then, the FLAI of risk j for period t is defined as

$$FLAI_{j,t} = F_{j,t}/M_j \tag{10.2}$$

$FLAI_{j.t}$ is 1 when $F_{j,t}$ is equal to M_j, and thus, the $FLAI_{j.t}$ of 1 denotes the general condition of foresight on risk j for the sample period T. The value of $FLAI_{j.t}$ greater than 1 denotes that for period *t*, it is foreseeable that risk j exposed by banks will become more severe than it is under general conditions. In contrast, for a $FLAI_{j.t}$ value of less than 1, losses caused by risk j foreseen for period t are considered to become less pronounced than the typical level of general condition.

10.2.4 Aggregate Bank Risks with the FLAI

As the last step, using the FLAI to adjust historical risk, we aggregate individual risks to obtain total risk. We first briefly describe how to measure historical risks based on historical numerical data recorded in financial statements, then we show how to adjust historical risks with the FLAI to obtain total risk.

Historical bank risks based on numerical data recorded in financial statements are derived as follows. Researchers can refer to Li et al. (2018) and Kretzschmar et al. (2010) for basic and specific background information. Specifically, the risk-return measure is written as

$$r_{i,j,t} = R_{i,j,t}/RWA_{i,t} \tag{10.3}$$

where $r_{i,j,t}$ and $R_{i,j,t}$ denote the risk-return and risk P&L of bank *i*, risk *j* for period *t*, respectively. $RWA_{i,r}$ denotes risk-weighted assets (RWA) of bank *i* for period *t*. The risk P&L data is collected by mapping risk types into income statement items.

The expected risk return of bank *i*, risk *j* for period T_i is written as

$$\overline{r_{i,j}} = \sum_{t=1}^{T_i} r_{i,j,t}/T_i \tag{10.4}$$

and the deviation from $r_{i,j,t}$ is written as

$$\Delta_{i,j,t} = r_{i,j,t} - \overline{r_{i,j}} \tag{10.5}$$

The risk-return of the typical bank *k* is written as

$$r_{k,j,t} = \overline{r_{k,j}} + \Delta_{i,j,t} \tag{10.6}$$

where $i = 1, 2, \ldots, n$. n is the number of banks in a banking system.

After obtaining the risk-return data for individual bank risks, Value-at-Risk (VaR), a popular and standard model for measuring and assessing risk, is used to measure individual risk in this chapter (Rosenberg and Schuermann 2006; Kretzschmar et al. 2010). However, VaR does not satisfy the subadditivity condition, so it is not a coherent risk measure (Artzner et al. 1999). A related statistic, expected shortfall (ES), which is also referred to as Conditional VaR, is a coherent risk measure that estimates the mean of the beyond VaR tail region (Rosenberg and Schuermann 2006). Thus, besides using VaR to discuss the empirical results in Sect. 10.5, we conduct robustness checks using ES as well.

VaR is defined as a quantile of the distribution of risk returns (Liu and Ralescu 2017). So the larger negative value or smaller positive value of VaR corresponds to the higher level of risk (Rosenberg and Schuermann 2006). VaR at a specific confidence level 1-α is defined as the smallest number l such that the probability of loss L exceeding l is not larger than α:

$$\text{VaR}(\alpha) = \inf\{l : P(L \geq l) \leq \alpha\} \tag{10.7}$$

Using (Eq. 10.7), ES at a specific confidence level 1-α is defined as the mean of the loss L exceeding VaR(α):

$$ES(\alpha) = E[L|L > VaR(\alpha)] \tag{10.8}$$

Similar to VaR, ES is taken as a tail expectation of losses so that ES is also negative.

To aggregate different risks into total risk, we adopt simple summation approach, which is one of the most basic risk aggregation approaches and has been widely used in many studies (Rosenberg and Schuermann 2006; Inanoglu and Jacobs 2009; Kretzschmar et al. 2010; Li et al. 2018). Simple summation, variance–covariance, and copula approaches are three commonly used risk aggregation approaches (Li et al. 2015). However, the variance–covariance and copula approaches are not applicable in this chapter. Specifically, the variance–covariance and copula approaches aggregate multiple risks by modeling the inter-risk correlations based on one-to-one data of different risks. However, by processing data using Eq. (10.3)–(10.6), the obtained data of credit, market and operational risks for the typical bank are not corresponding one by one. Thus, it is not practical to compute correlations between different risks based on the risk data of the typical bank. So the variance–covariance and copula risk aggregation approaches are not applicable in this chapter.

The simple summation approach assumes that all risks are perfectly correlated, i.e. that great losses occur simultaneously, which imposes an upper bound on the actual total risk (Dimakos and Aas 2004). Thus, the simple summation approach is found to be more conservative compared with other risk aggregation approaches (Embrechts et al. 1999). Obviously, the assumption that all risks are perfectly correlated is simple and unrealistic in most cases. Thus, the use of the simple summation approach only allows us to empirically test the impact of forward-looking information on total risk when all risk types coincide. The total risk value calculated based on historical numerical data recorded in financial statements under the 1–α confidence level for period t is written as:

$$Total - VaR_r(\alpha) = \sum_j VaR_{j,t}(\alpha) \tag{10.9}$$

where $VaR_{j,t}(\alpha)$ and $Total - VaR_r(\alpha)$ represent historical marginal risk j and historical total risk value, respectively.

After obtaining historical risks, forward-looking textual risk disclosures are incorporated into bank risk aggregation using $FLAI_{j,t}$ to adjust historical marginal risks. The forward-looking adjusted total risk value for period t is defined as

$$Adjusted - Total - VaR_t(\alpha) = \sum_j FLAI_{j,t^*} VaR_{j,t}(\alpha) \tag{10.10}$$

$FLAI_{j,t}$ has an amplifying or reducing effect on historical bank risk $VaR_{j,t}(\alpha)$. Specifically, $Total - VaR_t(\alpha)$ is a special case of $Adjusted - Total - VaR_t(\alpha)$ when $FLAI_{j,t}$ is equal to 1, denoting that the general condition of foresight on future bank risk movements is identical to past volatility characteristics, and no changes are brought about through the use of forward-looking information. Therefore, besides representing the general condition of foresight on future bank risk movements, $FLAI_{j,t}$ value equal to 1 is also assumed to represent the historical level of bank risks. Hence, the value of $FLAI_{j,t}$ greater than 1 increases risk j for period t because, according to forward-looking information, future risk j is considered to be more severe than historical values. Thus, historical risk values need to be extended to cover future risk losses. $FLAI_{j,t}$ value less than 1 reduces risk j for period t because forward-looking information projects future risk j to become less severe than historical levels. Hence, the FLAI can increase or reduce historical risk levels based on foresight on risks to attain a more reasonable total risk against future potential losses.

In summary, according to the above four steps, we incorporate forward-looking textual risk disclosures into bank risk aggregation. The total risk adjusted by forward-looking textual risk disclosures not only makes the total risk determined by historical risks but also reflects potential losses that banks may incur in the future. Thus, the forward-looking adjusted total risk is more reasonable and better able to cover potential total losses that may occur in the future, further guaranteeing the robust operation of banks.

10.3 Data Description

The backward-looking numerical data and forward-looking textual risk disclosures are collected from annual financial reports of U.S. commercial banks.

From the standard industrial classification (SIC) code list given on the SEC's website on commercial banks (Table 10.1), we obtained the names of U.S. commercial banks. The backward-looking numerical data of these U.S. commercial banks, including risk P&L and RWA data, were collected from the Thomson Reuters Eikon database.

Table 10.1 SIC of the commercial banking industry

	SIC	A/D office	Industry name
Commercial banks	6021	7	National commercial banks
	6022	7	State commercial banks
	6029	7	Commercial banks, NEC

Specifically, P&L arising from bank risks within specific periods is recorded and summarized in the income statement. Therefore, some researchers collected risk P&L data from income statements by mapping P&L items into risk types (Rosenberg and Schuermann, 2006; Kuritzkes and Schuermann, 2007; Inanoglu and Jacobs, 2009; Li et al., 2018; Zhu et al., 2018). The mapping relationships between income statement P&L items and different risk types are summarized in Table 10.2.

Based on the existing literature, we map credit risk into provisions, market risk into trading income, and operational risk into non-interest expenses to collect risk P&L data. Specifically, the provisions item in the income statement is a charge for incurred loan losses arising from credit default, which is inherently related to the definition of credit risk. A robustness check made by Kuritzkes and Schuermann (2007) found that the choice between provision and charge-offs appears to make little difference in measuring credit risk. Thus, in this chapter, we map credit risk into provisions as credit risk P&L data. Market risk refers to the potential losses arising from changes in the value or price of an asset, such as losses resulting from fluctuations in interest rates, currency exchange rates, stock prices, and commodity prices (BCBS 2006). In previous studies, taking trading income as the risk proxy has reached a consensus (Rosenberg and Schuermann, 2006; Kuritzkes and Schuermann, 2007; Inanoglu and Jacobs 2009). Therefore, in this chapter. we also map market risk into trading income.

Operational risk is defined as the risk of direct or indirect loss resulting from inadequate or failed internal processes, people and systems, or external events (BCBS 2006). The income statement item of non-interest expenses records expenses that occur outside of a bank's day-to-day activities, including costs from restructuring or reorganizing, penalty expenditure, compensatory payment, and abnormal losses. Thus, losses arising from operational risk events are mostly recorded under the item of non-interest expenses. Besides, Jacobs (2009) and Zhu et al. (2018) also mapped operational risk into non-interest expenses. Therefore, in this chapter, we map operational risk into non-interest expense to collect operational risk P&L data.

Table 10.2 The mapping between income statement P&L items and bank risks in previous studies

Studies	Mapping relationship		
	Credit risk	Market risk	Operational risk
Rosenberg and Schuermann (2006)	Net interest income - provisions	Trading revenue	——
Kuritzkes and Schuermann (2007)	-Provisions	Trading income	40% × (Other income + Non-interest expense + net extraordinary items)
Inanoglu and Jacobs (2009)	Gross charge-offs	Net trading revenues	Non-interest expense
Li et al. (2018)	Net interest income - provisions	Trading income	Remaining item
Zhu et al. (2018)	-provisions	Trading income	Non-interest expense

For the forward-looking textual risk disclosures, as of 2005, the SEC has required all filers to discuss company risk factors in an additional section (Sect. 1A) of Form 10-K. However, the effectiveness of risk disclosures has been widely questioned (Zhu et al., 2016). To enhance their effectiveness, the SEC issued comment letters asking filers only to list specific risk factors related to individual companies in 2010 (SEC 2010). Thus, to obtain more effective textual risk disclosures, we study the period from 2010 to 2017. Form 10-K filings were collected from EDGAR databases provided through the SEC's website.

Overall, we collected 812 pieces of numerical risk data for each credit, market, and operational risks and 36,178 summary headings of textual risk disclosures drawn from 1224 Form 10-K filings of 153 U.S. commercial banks for 2010–2017. To offer empirical insight into the total risk of U.S. commercial banks and make our results not specific to a real commercial bank, a typical U.S. commercial bank is constructed for our study. In accordance with Rosenberg and Schuermann (2006), Kretzschmar et al. (2010), and Alessandri and Drehmann (2010), we use medians to characterize this typical bank.

10.4 Empirical Results

In this section, we aggregate credit, market, and operational risks of the U.S. banking industry with forward-looking textual risk disclosures.

10.4.1 Forward-Looking Adjusted Total Risk

As discussed in Sect. 9.4 of Chap. 9, by using the semi-supervised text mining naive collision algorithm, we identify 21 bank risk factors: regulation, strategy, management operation, macroeconomic factor, loan loss, asset value fluctuation, capital availability, financial expense, competition, development of borrowers, reputation, product, and service, merger, and acquisition, accounting standard, financial institutions interaction, political environment, third party cooperation, liability obligation, disaster, country credit rating and other risk factors. The specific definitions of 21 bank risk factors are recorded in previous Table 9.3. The word clouds of 21 bank risk factors are shown in Fig. 9.3.

After identifying 21 bank risk factors, we map these 21 bank risk factors into credit, market, and operational risks based on definitions of bank risks and bank risk factors. The established mapping relationship is shown in Table 10.3. Specifically, credit risk is the risk of a loss caused by the failure of a counterparty to fulfill its contractual obligations (Perić et al. 2018). The corresponding risk factors include the development of borrowers and loan losses, which reflect the credit quality of borrowers and potential losses arising from defaults.

Table 10.3 The mapping relationship between risk factors and risk types

Risk type	Risk factor
Credit risk	Development of borrowers
	Loan loss
Market risk	Asset value fluctuation
	Capital availability
	Country credit rating
	Macroeconomic factor
	Financial institutions interaction
Operational risk	Competition
	Financial expense
	Liability obligation
	Management operation
	Strategy
	Political environment
	Reputation
	Product and service
	Accounting standard
	Merger and acquisition
	Third-party cooperation
	Regulation
	Disaster
	Other risk factors

Market risk measures the loss due to declines in the value of financial positions held by banks, which is generally determined based upon market prices, such as interest rate and exchange rate (Hartmann 2010). So in this chapter, the risk factors that will cause declines in the value of financial positions held by banks are mapped into market risk, including asset value fluctuation, capital availability, country credit rating, macroeconomic factor, and financial institutions interaction. To be specific, asset value fluctuation directly describes that declines in the value of securities held in the investment portfolio may negatively affect the bank's earnings and capital. Capital availability describes the loss arising from failing to fund capital by liquidating securities at a fair value. The deterioration of country credit rating will cause bank loss by reducing the market value of treasury and government agency securities held by banks. The financial institution's interaction describes that banks in crisis will sell securities to support liquidity, which further causes declines in the value of securities held by other banks. The fluctuations of macroeconomic factors, such as the interest rate and exchange rate will cause declines in the value of the banks' investment portfolios. Thus, asset value fluctuation, capital availability, country credit rating, macroeconomic factor, and financial institutions interaction are five market risk factors.

Basel II Accord defines operational risk as the loss resulting from inadequate or failed internal processes, people and systems, or external events (BCBS 2006; Feria-Domínguez et al. 2015). Actually, operational risk in the banking industry started life as a residual category for those risks and uncertainties that cannot be categorized into credit or market risks (Power 2005; Li et al. 2018). So in this chapter, the residual 14 bank risk factors are mapped into operational risk: competition, financial expense, liability obligation, management operation, strategy, political environment, reputation, product and service, accounting standard, merger and acquisition, third party cooperation, regulation, disaster, and other risk factors. However, the mapping relationship between risk factors and operational risk is kind of rough. Overall, although the mapping relationship is hardly perfect, we believe it still provides a reasonable basis for linking risk factors into different risk types, allowing us to incorporate forward-looking textual risk disclosures into bank risk aggregation.

By mapping credit, market, and operational risks into 21 bank risk factors, we obtain word clouds for credit, market, and operational risks, where the font size corresponds to the probability of the word occurring in the risk type (Fig. 10.2). Thus, the larger the size of the word, the more frequent the word appears in describing the risk type. It is clear from Fig. 10.2 that the high-frequency words of each risk type capture the essential characteristics of credit, market, and operational risks, showing that the mapping relationship between risk types and bank risk factors is reasonable and acceptable.

Specifically, the high-frequency words for credit risk are loan and credit, representing the nature of credit risk: the credit quality of borrowers. Regarding market risk, high-frequency words include interest, market, stock, and economy, which reflect that market risk loss is caused by changes in financial markets and macroeconomic conditions. As for operational risk, high-frequency words include business, regulation, and operation, which describe the business operating losses due to internal processes, people and systems, or external events (BCBS 2006).

Then, based on annual disclosure frequencies for credit, market, and operational risks recorded in Table 10.4, we calculate annual FLAIs for credit, market, and operational risks for 2010–2017 (Table 10.4). Specifically, for 2010–2017, the annual disclosure frequencies of credit, market, and operational risks vary from 408 to 679, 1112 to 1276, and 2312 to 3260, respectively.

Fig. 10.2 Word clouds for credit, market, and operational risks

Table 10.4 The annual disclosure frequency and FLAI of credit, market, and operational risks

Risk year	Credit risk		Market risk		Operational risk	
	Frequency	FLAI	Frequency	FLAI	Frequency	FLAI
2010	379	0.81	1102	0.90	2351	0.83
2011	421	0.90	1174	0.96	2616	0.92
2012	475	1.02	1292	1.06	2589	0.91
2013	442	0.95	1233	1.01	2726	0.96
2014	522	1.12	1279	1.05	2894	1.02
2015	590	1.27	1268	1.04	2944	1.04
2016	452	0.97	1192	0.98	3257	1.15
2017	443	0.95	1217	1.00	3320	1.17

Furthermore, the annual FLAIs for credit, market, and operational risks from 2010 to 2017 fluctuate at approximately 1 as shown in Fig. 10.3. In terms of fluctuations, the FLAI for credit risk increased significantly in 2014 and 2015, while market risk FLAI remained relatively minor. The FLAI for operational risk follows an upward trend. Regarding extreme values, the smallest FLAIs for credit, market, and operational risks are measured for 2010 as 0.81, 0.90, and 0.83, respectively. The largest FLAI for credit risk was 1.27, recorded for 2015. Regarding market risk, the largest FLAI was recorded for 2012 at 1.06. With an upward tendency, the largest FLAI for operational risk of 1.17 was recorded for 2017.

Having obtained the annual FLAI values of credit, market, and operational risks, we then measure historical bank risk based on 812 pieces of numerical risk data for each credit, market, and operational risks from 2010 to 2017. The marginal risk distributions of credit, market, and operational risks, whose distributional shapes vary considerably, are illustrated in Fig. 10.4. The specific characteristics of marginal risk distributions are presented in Table 10.5.

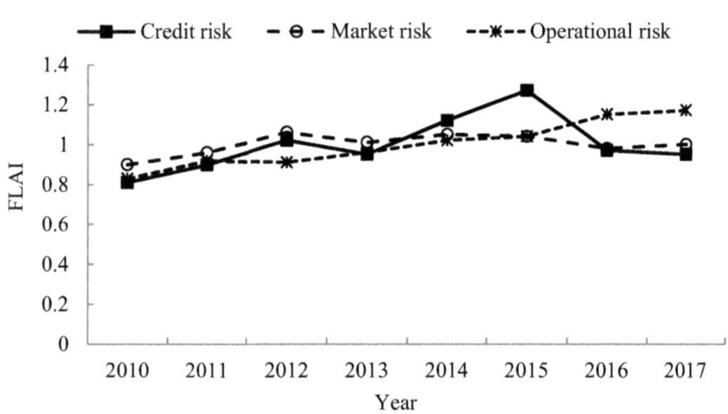

Fig. 10.3 The annual FLAI for credit, market, and operational risks from 2010 to 2017

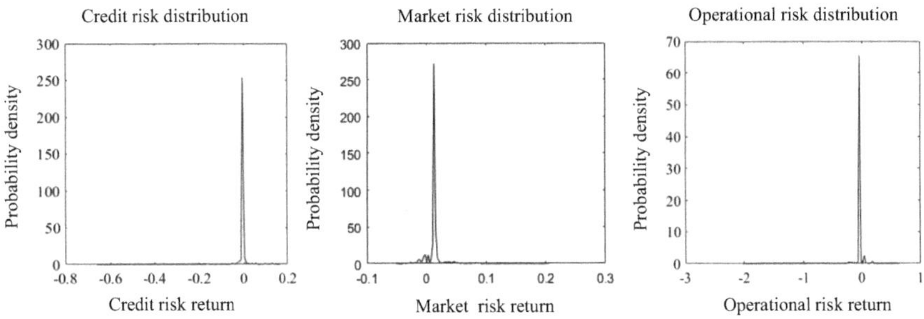

Fig. 10.4 Marginal risk distributions of credit, market, and operational risks

Table 10.5 Descriptive statistics for credit, market, and operational risk distributions

	Credit risk	Market risk	Operational risk
$\mu(\%)$	−0.51	1.31	-4.62
$\sigma(\%)$	3.67	1.58	10.34
Skewness	−9.66	2.99	-8.96
Kurtosis	155.06	32.43	180.74

Specifically, in terms of risk return on RWA, operational risk has the highest volatility (10.34%), the highest tails (kurtosis = 180.74), and the smallest mean value (−4.62%). Generally, operational risk distribution is considered to be fat-tails due to extreme loss events (Panjer 2006; Zhu et al. 2019). The kurtosis values for credit risk and market risk are 155.06 and 32.43, respectively, with their volatilities measured as 3.67% and 1.58%, respectively. The mean value of credit risk is negative (−0.51%), while the mean value of market risk is positive (1.31%). Credit risk is the most significantly left-skewed (Skewness = −9.66), and operational risk is left-skewed at −8.96. The skewness of market risk is equal to 2.99, which means market risk is right-skewed.

Historical VaR values of individual risk measured at the 99% confidence level for different years are recorded in Table 10.6 For example, for 2017, at the 99% confidence level, the VaR values of credit, market, and operational risks are recorded as −8.38%, −3.16% and −26.33%, respectively.

Table 10.6 VaR values for credit, market and operational risks from 2010 to 2017 at the 99% confidence level/%

Year risk	2010	2011	2012	2013	2014	2015	2016	2017
Credit risk	−8.93	−9.10	−8.87	−8.66	−8.54	−8.46	−8.41	−8.38
Market risk	−0.82	−0.79	−0.93	−0.68	−1.44	−2.56	−2.71	−3.16
Operational risk	−23.32	−24.18	−27.05	−26.97	−26.83	−26.61	−26.44	−26.33

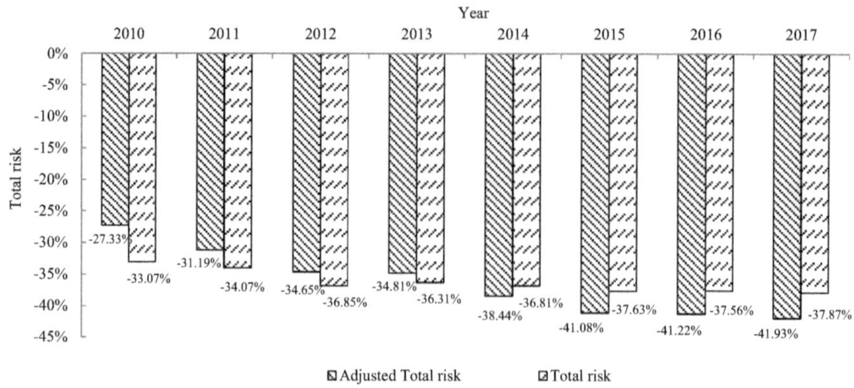

Fig. 10.5 Comparison of total risk with and without textual risk disclosure from 2010 to 2017

In the following, we use forward-looking textual risk disclosures to adjust the historical bank risk to determine forward-looking adjusted total risk. Specific VaR values of the adjusted total risk of the 99% confidence level are shown in Fig. 10.5. Specifically, from 2010 to 2017, the lowest adjusted total risk level with VaR being −27.33% is recorded for 2010. The highest level of adjusted total risk is −41.93%, which indicates that U. S. commercial banks faced the most severe risks in 2017.

Values of total risk without forward-looking textual risk disclosures at 99% confidence level for 2010–2017 are shown in Fig. 10.5 for comparative purposes. We also calculate the change rate to quantify the change brought about by forward-looking textual risk disclosures. The change rate is the proportional change in total risk using total risk value with forward-looking textual risk disclosures $(Adjusted\,total\,risk)$ versus total risk value without forward-looking textual risk disclosures $(Total\,risk)$: $(Adjusted\,total\,risk - Total\,risk)/Total\,risk$ (Fig. 10.6). The positive change rate value illustrates that the inclusion of forward-looking textual risk disclosures in bank risk aggregation amplifies total risk. In contrast, the negative value of the change rate shows that incorporating forward-looking textual risk disclosures into bank risk aggregation has a reducing effect on total risk.

According to Figs. 10.5 and 10.6, it is clear that from 2010 to 2013, the adjusted total risk was less than the total risk, which leads the change rate to be negative, showing that forward-looking textual risk disclosures have a reducing effect on total risk. Disregarding forward-looking textual risk disclosures in bank risk aggregation will overestimate total risk by −17.37%, −8.44%, −5.97%, and −4.14%, respectively. The reason for this case may be that banks have not yet recovered from the subprime crisis that began in 2007. Thereby the severe operating condition is recorded in financial statements with numerical risk data. However, the Federal Reserve's quantitative easing (QE) policies implemented in the wake of the subprime crisis have positively affected financial markets, and bankers anticipate that QE policies may mitigate bank risks faced (Joyce et al. 2012). Thus, when forward-looking

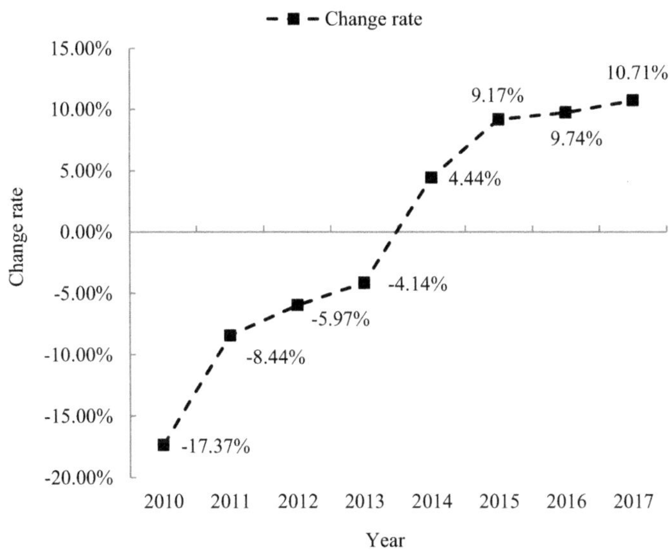

Fig. 10.6 Change rates brought about by forward-looking textual risk disclosures for 2010–2017

textual risk disclosures cover positive signals for future markets, taking them into risk aggregation will lower the total risk.

In contrast, for 2014–2017, the adjusted total risk was higher than the total risk, resulting in a positive change rate. Thus, forward-looking textual risk disclosures amplify the total risk, while disregarding forward-looking textual risk disclosures in bank risk aggregation underestimates the total risk by 4.44%, 9.17%, 9.74%, and 10.71%, respectively. This situation may be the case that the extended downturn in the global economy subjected the banking industry to difficulties. Negative profits for 2015 disclosed by Deutsche Bank, the largest bank in Europe, and the British chartered group caused bank shares to plunge in Europe and the U.S. On February 21, 2015, Goldman Sachs released a report on the performance of US banks, in which Goldman Sachs admits that after the gloomy 2015, the U.S. banking sector would face a bad start in the first quarter of 2016 (Goldman Sachs 2015). Thus, bankers are not optimistic about the future and believe that more severe losses may occur. Thus, considering negative forward-looking textual risk disclosures increases total risk. To summarize, disregarding forward-looking textual risk disclosures during bank risk aggregation overestimates or underestimates total risk.

10.4.2 Robust Test of Total Risk

Since VaR is not a coherent risk measure based on the criteria of Artzner et al. (1999), we repeat the experiments using the ES measure as defined in (Eq. 10.8).

Table 10.7 ES values for credit, market and operational risks from 2010 to 2017 at the 99% confidence level/%

Year risk	2010	2011	2012	2013	2014	2015	2016	2017
Credit risk	−21.52	−23.94	−27.39	−27.18	−27.05	−26.98	−26.93	−26.89
Market risk	−1.52	−1.50	−1.22	−1.48	−2.63	−3.51	−3.74	−4.20
Operational risk	−56.80	−59.05	−60.75	−61.10	−60.96	−60.74	−60.56	−60.45

The general conclusion implied by VaR holds. Specifically, historical ES values of individual risk measured at the 99% confidence level for different years are recorded in Table 10.7. For example, in 2017, the ES values of credit, market, and operational risks are −26.89%, −4.20%, and −60.45%, respectively. The ES values of credit and market risks are far less than their VaR values, indicating that there are extreme losses beyond the VaR tail region, which is in line with the fat-tail characteristics of credit and operational risk distributions. Table 10.5 records that operational risk has the highest fattest tails (kurtosis = 180.74). And credit risk is also characterized by fat tails with a kurtosis value being 155.06. Compared with operational and credit risks, market risk has the thinnest tails (kurtosis = 32.43).

Table 10.8 records the total risk with and without forward-looking textual risk disclosures and the change rate caused by incorporating textual risk disclosures into bank risk aggregation. It is clear from Table 10.8 that the general conclusions are no different from our previous findings using VaR. To be specific, the change rates from 2010 to 2013 are negative, while from 2014 to 2017 are positive. This is analogous to the VaR results displayed in Fig. 10.6. Similarly, the change rates calculated based on both VaR and ES are very close and range from -18% to 11%. Specifically, from 2010 to 2013, the adjusted total risk was less than the total risk, showing that forward-looking textual risk disclosures have a reducing effect on total risk. Disregarding forward-looking textual risk disclosures in bank risk aggregation will overestimate total risk by −17.41%, −8.50%, −5.42%, and −4.22%, respectively. In contrast, the adjusted total risk is greater than the total risk from 2014–2017. Thus, disregarding forward-looking textual risk disclosures in bank risk aggregation underestimates the

Table 10.8 ES values for total risk with and without textual risk disclosures from 2010 to 2017 at the 99% confidence level/%

Risk year	Adjusted total risk	Total risk	Change rate
2010	−65.94	−79.84	−17.41
2011	−77.31	−84.49	−8.50
2012	−84.51	−89.36	−5.42
2013	−85.97	−89.76	−4.22
2014	−95.24	−90.64	5.07
2015	−101.09	−91.23	10.80
2016	−99.43	−91.23	8.99
2017	−100.47	−91.54	9.76

total risk by 5.07%, 10.8%, 8.99%, and 9.76%, respectively. Overall, the robustness check using ES also shows that disregarding forward-looking textual risk disclosures overestimates the total risk of 2010–2013 while underestimating the total risk of 2014 to 2017.

10.5 Conclusions

This chapter innovatively incorporates Forward-looking textual risk disclosures reported in financial statements into bank risk aggregation to attain more reasonable values of total risk. Bank risk aggregations in previous studies have mainly been based on historical numerical data recorded in financial statements. However, risk losses that have not occurred in the past may occur in the future. Some risk losses can become more severe with changing market conditions. Thus, only using historical risk data may lead to biased aggregation results. The use of forward-looking information that foresees potential future risk losses is essential to bank risk aggregation.

Based on 812 pieces of numerical risk data on credit, market, and operational risks and 36,178 summary headings of textual risk disclosures of 1224 Form 10-K filings of 153 U.S. commercial banks for 2010–2017, we aggregate credit, market, and operational risks of U.S. commercial banks. Most of the discussion is in terms of VaR; we conduct robustness checks using ES and find that all results implied by VaR hold. Specifically, by comparing total risk VaR with and without forward-looking textual risk disclosures, we found that ignoring forward-looking textual risk disclosures overestimates total risk by −17.37%, −8.44%, −5.97%, and −4.14% from 2010 to 2013 while underestimating total risk by 4.44%, 9.17%, 9.74%, and 10.71% from 2014 to 2017. Thus, disregarding forward-looking textual risk disclosures in bank risk aggregation leads to the overestimated or underestimated total risk.

This chapter is not without limitations. The limitation is that the mapping relationship between risk factors and risk types is hardly perfect. Mapping the residual risk factors that cannot be categorized into credit and market risks to operational risk is kind of rough. In future research, the inclusion of other information may help to calibrate the mapping relationship to some extent.

References

Abed S, Al-Najjar B, Roberts C (2016) Measuring annual report narratives disclosure: empirical evidence from forward-looking information in the UK prior the financial crisis. Manag Audit J 31(4/5):338–361. https://doi.org/10.1109/ICSC.1996.864276

Alessandri P, Drehmann M (2010) An economic capital model integrating credit and interest rate risk in the banking book. J Bank Finance 34(4):730–742. https://doi.org/10.1016/j.jbankfin.2009.06.012

Artzner P, Delbaen F, Eber JM, Heath D (1999) Coherent measures of risk. Math Finance 9(3):203–228. https://doi.org/10.1111/1467-9965.00068

Athanasakou V, Hussainey K (2014) The perceived credibility of forward-looking performance disclosures. Account Bus Res 44(3):227–259. https://doi.org/10.1080/00014788.2013.867403

Bao Y, Datta A (2014) Simultaneously discovering and quantifying risk types from textual risk disclosures. Manage Sci 60(6):1371–1391. https://doi.org/10.1287/mnsc.2014.1930

Basel Committee on Banking Supervision (2006) International convergence of capital measurement and capital standards: a revised framework. Bank for International Settlements, Basel

Baselga-Pascual L, Trujillo-Ponce A, Cardone-Riportella C (2015) Factors influencing bank risk in Europe: evidence from the financial crisis. The North Am J Econ Finance 34:138–166. https://doi.org/10.1016/j.najef.2015.08.004

Beneish MD, Miller BP, Yohn TL (2015) Macroeconomic evidence on the impact of mandatory IFRS adoption on equity and debt markets. J Account Public Policy 34(1):1–27. https://doi.org/10.1016/j.jaccpubpol.2014.10.002

Breuer T, Jandacka M, Rheinberger K, Summer M (2010) Does adding up of economic capital for market and credit risk amount always to conservative risk estimates? J Bank Finance 34(4):703–712. https://doi.org/10.1016/j.jbankfin.2009.03.013

Campbell JL, Chen HC, Dhaliwal DS, Lu HM, Steele LB (2014) The information content of mandatory risk factor disclosures in corporate filings. Rev Acc Stud 19(1):396–455. https://doi.org/10.1007/s11142-013-9258-3

Dimakos XK, Aas K (2004) Integrated risk modelling. Stat Model 4(4):265–277. https://doi.org/10.1191/1471082X04st079oa

Drehmann M, Sorensen S, Stringa M (2010) The integrated impact of credit and interest rate risk on banks: a dynamic framework and stress testing application. J Bank Finance 34(4):713–729. https://doi.org/10.1016/j.jbankfin.2009.06.009

Dyer T, Lang M, Stice-Lawrence L (2017) The evolution of 10-K textual disclosure: evidence from latent dirichlet allocation. J Account Econ 64(2–3):221–245. https://doi.org/10.1016/j.jacceco.2017.07.002

Embrechts P, McNeil AJ, Straumann D (1999) Correlation: pitfalls and alternatives. Risk 12:69–71

Feria-Domínguez JM, Jimenez-Rodriguez E, Sholarin O (2015) Tackling the over-dispersion of operational risk: implications on capital adequacy requirements. The North Am J Econ Finance 31:206–221. https://doi.org/10.1016/j.najef.2014.11.004

Feuerriegel S, Gordon J (2018) Long-term stock index forecasting based on text mining of regulatory disclosures. DecisSupport Syst 112:88–97. https://doi.org/10.1016/j.dss.2018.06.008

Goldman Sachs (2015) Goldman Sachs Bank USA and Subsidiaries-2015 Annual Report. Retrieved from http://www.goldmansachs.com/investor-relations/financials/archived/subsidiary-financial-info/gsbank-usa/gs-bank-usa-2015-financial-statements.pdf

Grundke P (2010) Top-down approaches for integrated risk management: How accurate are they? Eur J Oper Res 203(3):662–672. https://doi.org/10.1016/j.ejor.2009.09.015

Hartmann P (2010) Interaction of market and credit risk. J Bank Finance 34(4):697–702. https://doi.org/10.1016/j.jbankfin.2009.10.013

Huang KW, Li ZL (2011) A multilabel text classification algorithm for labeling risk factors in SEC form 10-K. ACM Trans Manage Inf Syst (TMIS) 2(3):1–19. https://doi.org/10.1145/2019618.2019624

Inanoglu H, Jacobs M (2009) Models for risk aggregation and sensitivity analysis: an application to bank economic capital. J Risk Finan Manage 2(1):118–189. https://doi.org/10.3390/jrfm2010118

Jarrow RA, Turnbull SM (2000) The intersection of market and credit risk. J Bank Finan 24(1):271–299. https://doi.org/10.1016/S0378-4266(99)00060-6

Jiménez G, Lopez JA, Saurina J (2013) How does competition affect bank risk-taking? J Financ Stab 9(2):185–195. https://doi.org/10.1016/j.jfs.2013.02.004

Joyce M, Miles D, Scott A, Vayanos D (2012) Quantitative easing and unconventional monetary policy–an introduction. Econ J 122(564):271–288. https://doi.org/10.1111/j.1468-0297.2012.02551.x

Kılıç M, Kuzey C (2018) Determinants of forward-looking disclosures in integrated reporting. Manag Audit J 33(1):115–144. https://doi.org/10.1108/MAJ-12-2016-1498

Kretzschmar G, McNeil AJ, Kirchner A (2010) Integrated models of capital adequacy-why banks are undercapitalized. J Bank Finance 34(12):2838–2850. https://doi.org/10.1016/j.jbankfin.2010.02.028

Kupiec PH, Ramirez CD (2013) Bank failures and the cost of systemic risk: evidence from 1900 to 1930. J Finan Intermed 22(3):285–307. https://doi.org/10.1016/j.jfi.2012.09.005

Kuritzkes A, Schuermann T (2007) What we know, don't know and can't know about bank risk: a view from the trenches. In Diebold FX, Doherty NA, Herring RJ (eds) The known the unknown and the unknowable in financial risk management. Princeton University Press, Princeton

Li J, Zhu X, Lee CF, Wu D, Feng J, Shi Y (2015) On the aggregation of credit, market and operational risks. Rev Quant Financ Acc 44(1):161–189. https://doi.org/10.1007/s11156-013-0426-0

Li J, Wei L, Lee CF, Zhu X, Wu D (2018) Financial statements based bank risk aggregation. Rev Quant Financ Acc 50(3):673–694. https://doi.org/10.1007/s11156-017-0642-0

Liu Y, Ralescu DA (2017) Value-at-risk in uncertain random risk analysis. Inf Sci 391:1–8. https://doi.org/10.1016/j.ins.2017.01.034

Lüdering J, Tillmann P (2018) Monetary policy on twitter and asset prices: evidence from computational text analysis. The North Am J Econ Finan. In press. https://doi.org/10.1016/j.najef.2018.11.004

Miller GS (2017) Discussion of "the evolution of 10-K textual disclosure: evidence from latent dirichlet allocation." J Account Econ 64(2–3):246–252. https://doi.org/10.1016/j.jacceco.2017.07.004

Muslu V, Radhakrishnan S, Subramanyam KR, Lim D (2014) Forward-looking MD&A disclosures and the information environment. Manage Sci 61(5):931–948. https://doi.org/10.1287/mnsc.2014.1921

Panjer HH (2006) Operational risk: modeling analytics. Wiley

Perić BŠ, Smiljanić AR, Aljinović Z (2018) Credit risk of subsidiaries of foreign banks in CEE countries: impacts of the parent bank and home country economic environment. The North Am J Econ Finan 46:49–69. https://doi.org/10.1016/j.najef.2018.03.009

Power M (2005) The invention of operational risk. Rev Int Polit Econ 12(4):577–599. https://doi.org/10.1080/09692290500240271

Rosenberg JV, Schuermann TA (2006) General approach to integrated risk management with skewed, fat-tailed risks. J Financ Econ 79(3):569–614. https://doi.org/10.1016/j.jfineco.2005.03.001

Securities and Exchange Commission (SEC) (2005) Securities and exchange commission final rule, release no. 33–8591(FR-75). Retrieved from http://www.sec.gov/rules/final/33-8591.pdf

Securities and Exchange Commission (SEC) (2010) Annual report pursuant to section 13 or 15(d) of the securities exchange act of 1934. Retrieved from http://www.sec.gov/about/forms/form10-k.pdf

Wei L, Li G, Zhu X, Li J (2019a) Discovering bank risk factors from financial statements based on a new semi-supervised text mining algorithm. Account Finan. In press. https://doi.org/10.1111/acfi.12453

Wei L, Li G, Zhu X, Sun X, Li J (2019b) Developing a hierarchical system for energy corporate risk factors based on textual risk disclosures. Energy Econ 80:452–460. https://doi.org/10.1016/j.eneco.2019.01.020

Zhu X (2017) Semi-supervised learning literature survey. Comput Sci 37(1):63–77

Zhu X, Wand Y, Li J (2019) Operational risk measurement: a loss distribution approach with segmented dependence. J Oper Risk 14(1):1–20. https://doi.org/10.21314/JOP.2019.220

Zhu X, Wei L, Wu D, Li J (2018) A general framework for constructing bank risk data sets. J Risk 21(1):37–59. https://doi.org/10.21314/JOR.2018.393

Zhu X, Yang SY, Moazeni S (2016) Firm risk identification through topic analysis of textual financial disclosures. Comput Intell IEEE, 1-8. https://doi.org/10.1109/SSCI.2016.7850005

Chapter 11
Main Conclusions and Future Research

11.1 The Main Research Work

At present, the related researches have solved the problem of insufficient data by establishing the mapping relationship between the income statement, balance sheet, and bank risks, and collecting risk data from financial statements. However, there are three key problems in the existing researches on bank risk integration using financial statement data: low frequency of financial statement data, neglect of off-balance sheet business risk, lag of financial statement data, which may lead to inaccurate results of bank risk integration. To solve these three key problems, further advance the researches of bank risk integration based on financial statements, and improve the rationality and reliability of bank risk integration results, the main works of this book are as follows:

To solve the key problem of low-frequency data, this book proposes a factor integral risk integration method based on high-frequency risk factors. Although the frequency of risk data is low (usually quarterly), the frequency of potential risk factor data is high (usually daily data). Therefore, the factor integration method solves the key problem of low-frequency risk data by transforming the integration of low-frequency risk data into the integration of high-frequency risk factors. Rich historical data based on high-frequency risk factors can improve the rationality and robustness of bank risk integration results.

Regarding the problem of neglecting the off-balance sheet business risk, this book establishes the mapping relationship between the off-balance sheet business and the bank risk types, integrates the off-balance sheet business into the bank risk integration, measures the risks brought by the off-balance-sheet business comprehensively, and solves the problem that the risk integration results are biased due to the neglect of off-balance sheet business in previous studies.

As for the key problem of data lag, this book proposes a risk integration method combining historical data with forward-looking text risk information. To analyze forward-looking risk text information more accurately, a new semi-supervised naive collision text mining algorithm is proposed. At the same time, the historical data and

J. Li et al., *Financial Statements-Based Bank Risk Aggregation*, Innovation in Risk Analysis, https://doi.org/10.1007/978-981-19-0408-0_11

forward-looking risk information in financial statements are used in risk integration. The risk integration results not only reflect the historical situation of risk but also consider the future trend of risk, which increases the timeliness of risk integration results and can estimate the potential loss in the future more reasonably.

11.2 The Main Conclusions

This book uses the financial statement data of China's listed banks and American banks to conduct the empirical studies. there are three major conclusions.

(1) The results of bank risk integration based on low-frequency risk data will underestimate the overall risk faced by banks. Through the factor integration method, this paper transforms the integration of low-frequency risk data into the integration of high-frequency risk factors, which improves the rationality and robustness of bank risk integration results. In the part of empirical research, based on the financial statements of 16 listed banks in China, the credit risk and market risk of Chinese listed banks are integrated by the factor integral method. The empirical results show that the overall risk distribution based on low-frequency risk profit and loss data is quite different from that based on high-frequency risk factors. In addition, compared with the economic capital estimation based on low-frequency risk profit and loss data, the economic capital obtained by the factor integration method proposed in this paper is about 21% higher, which indicates that banks need to hold more capital to prevent unexpected losses caused by credit and market risks.

(2) In the bank risk integration, ignoring the off-balance sheet business will overestimate the risk of large commercial banks and underestimate the risk of small commercial banks. Based on the financial statement data of 16 listed banks in China, the empirical research part integrates the credit risk, market risk, liquidity risk, and operational risk of China's banking industry. By constructing two hypothetical banks with different scales, this paper further studies whether the impact of off-balance sheet business on banks' overall risk is related to the size of banks. The experimental results show that the off-balance sheet business reduces the overall risk of large commercial banks and increases the overall risk of small commercial banks. In addition, with the further development of off-balance sheet business and the gradual improvement of information disclosure of off-balance sheet business, the deviation of risk integration result caused by ignoring off-balance sheet business will be larger and larger.

(3) The timeliness of risk integration results based on historical data is poor. The risk integration results are lagged because the future trend of risk is ignored in the risk data used in the bank risk integration. Therefore, the introduction of future trends can offset the lag. In the part of empirical research, this paper integrates the credit risk, market risk, and operational risk of the American banking industry based on the financial statement data of 153 American commercial

banks. The experimental results show that the semi-supervised naive collision algorithm proposed in this paper achieves an average classification accuracy of three times compared with the unsupervised text mining algorithm. In the process of risk integration, ignoring the future trend of risk will overestimate the overall risk in 2010–2013 and underestimate the overall risk in 2014–2017. Therefore, it is necessary to incorporate forward-looking risk information into risk integration to enhance the timeliness of risk integration results, to more accurately deal with potential losses that may occur in the future and maintain the stable operation of the bank.

11.3 Future Researches

Based on the in-depth study of bank risk integration based on financial statements, this book solves three critical problems, improves the research of bank risk integration based on financial statements, and obtains more reasonable and reliable results of bank risk integration. In the following, we will propose four directions for perfecting researches:

11.3.1 Accurate Mapping Relationship

This book establishes the mapping relationship between risk types and income statement, assets on and off-balance sheet, and risk factors. However, there are still some defects in these mapping relationships. For example, the net interest income of the income statement is affected not only by market risk but also by credit risk. The risk factor "the connection between financial institutions" affects not only the market risk but also the credit risk. In addition, this book classifies the residual items and residual risk factors that can not be mapped to credit risk and market risk into operational risk. Therefore, our next research direction can use other information, such as risk information disclosed in credit rating reports and analyst reports to further refine the mapping relationship between risk types and financial statements.

11.3.2 A "Three Statements in One" Bank Risk Aggregation Approach

Income statement, balance sheet, and cash flow statement are the three main financial statements, which reflect the bank's historical operation status. However, the current research on bank risk integration based on financial statements only uses the information in the income statement and the balance sheet, but not the data in the cash flow statement. In future research, the cash flow statement should also be incorporated into the bank risk integration because it also played a very important role in the further refinement of the mapping relationship between risk types and financial statements. The Mapping relationship between risk types and cash flow statementwill make the framework of financial statement based bank risk aggregation more comprehensive and reasonble. A "three statements in one" bank risk aggregation approach can be proposed to more accurately aggregate bank risks.

11.3.3 Full Coverage of Bank Risk Types

The existing researches on bank risk integration based on financial statements only integrate the main risk types of banks, including credit risk, market risk, operational risk, and liquidity risk. However, there are other types of risk in the bank's operation, such as the reputation risk, legal risk, country risk and strategic risk. With the increasing impact of the four risks on banks, these risks will be further incorporated into the bank risk integration based on financial statements in future research, to describe the overall risk of banks more comprehensively.

11.3.4 Incorporate Multiple-Source Financial Texts

From the perspective of big data, the data sources of bank risk can be further expanded. In addition to the financial statements recording bank's risk information, there are many other data sources containing a large number of bank risk information, such as news reports, credit rating reports, regulatory reports, and investment analysts' analysis reports, which also contain abundant information about bank risks. In the future, on the basis of collecting bank risk information from financial statements, this paper will further collect bank risk information from other multiple-source financial texts through text mining technology to describe bank risks faced more accurately.

Printed by Printforce, the Netherlands